# Hermeneutics at the Intersection of Medical Technology

Jan Kyrre Berg Olsen Friis

# Hermeneutics at the Intersection of Medical Technology

## Interpretation Reimagined

palgrave
macmillan

Jan Kyrre Berg Olsen Friis
Public Health Science
University of Copenhagen
Copenhagen, Denmark

ISBN 978-3-031-88754-3 ISBN 978-3-031-88755-0 (eBook)
https://doi.org/10.1007/978-3-031-88755-0



This Palgrave Macmillan imprint is published by the registered company Springer Nature Switzerland AG.
The registered company address is: Gewerbestrasse 11, 6330 Cham, Switzerland

*To my dearest loved ones. To my beautiful wife, Teresa, to my cherished daughter, Mathilde, and my wonderful son-in-law, Mason, to my precious granddaughters, Eleanor and Nathalie, to my beloved son, Henrik, and my dear daughter-in-law, Delia, and with joy in our hearts, to my soon-to-be-born grandson, Alexander.*

# Competing Interests

The author has no competing interests to declare that are relevant to the content of this manuscript.

# About the Book

*Interpretation Reimagined* challenges the traditional view of radiology as a purely technical, objective field by emphasizing the interpretative and embodied nature of diagnostic practice. Rather than functioning as transparent, mechanical processes, knowing, perceiving, and acting form an intricate web, rooted in the dynamic relationship between the body, mind, and environment. Much of cognition operates tacitly, shaping expertise through implicit knowledge that enables seamless interpretation of complex data. In collaborative medical settings, truth emerges not as a fixed endpoint but through shared inquiry, adapting fluidly to each patient's unique condition.

Cognitive biases, often seen as limitations, may serve as essential interpretative shortcuts that shape perception and decision-making. Memory, a continuously evolving and reconstructive process, links past experiences to present contexts, guiding clinicians as they navigate uncertainty. This dynamic interplay—where knowledge is fluid, embodied, and adaptive—reshapes our understanding of expertise. Rather than striving for absolute certainty, medical practice depends on the ability to interpret, revise, and refine knowledge in response to new insights. Ultimately, this perspective highlights interpretation as central to clinical reasoning, reinforcing the necessity of flexibility, collaboration, and embodied intelligence in navigating the complexities of modern medicine.

* * *

***Light within the Abyss***
At the heart of human experience, a thought self-born resides
A silent spring of understanding where reason and emotion collide.
From this merging flows the essence of all that we create
A spark of unity, the force that shapes our fate.
Reason, a humble tool, a lens to seek and see
Neither good nor evil – just what it strives to be.
An instrument of clarity, a servant to the mind
Yet aimless, adrift, unless a purpose it can find.
Here, intuition whispers, a dreamer's guiding hand
A flicker of imagination, like starlight over land.
For without the heart's deep current, the soul's enduring call
Reason fades to an empty shell, a tool and nothing more at all.

# Contents

CHAPTER 1

# Hermeneutics and Medical Practice

An aging Martin Heidegger reemphasized in his 1966 book *Discourse on Thinking* the tension between two quite different modes of thinking—both typical of the human mind and intellect. One mode he termed "calculative thinking," which is a conscious, intentional, and methodical deliberation of consequences and probabilities, and of opportunities. This thinking calculates utility by estimating the usability, cost, and value of things. Moreover, calculative thinking is, according to Heidegger, an objectifying thinking and as such, an intended disinterested pursuit of facts. It is a style of thinking that, as John D. Caputo puts it, and in agreement with Heidegger, characterizes the mathematical-technical sciences and modern technical philosophy.[1] It is the thinking typical of our age. "For the present age stands dedicated to a technical understanding of reality and to the control of reality which this makes possible," Caputo writes.[2] The abstracting ability of mind is also the ability to take a step back from the concreteness of affairs with disinterest. The intellectual philosopher, physicist, or medical clinician will nevertheless have to struggle in order to keep their disinterested and detached gaze. The hurdle is to detach from personal, group, or organizational motives, interests, and needs defining the specific context of philosophical research or scientific practice. The rational philosopher, physician, or engineer, emphasizing calculative

[1] Caputo (1986: 60).
[2] Ibid.



J. K. B. O. Friis, *Hermeneutics at the Intersection of Medical Technology*, https://doi.org/10.1007/978-3-031-88755-0_1

thinking, seems to share an assumption that one can ***subdue*** the influence of "subjective inputs" if one immerses oneself in the ***deliberative process*** of critical examination and subsequent exchange of information in order to inform their agreed-upon decisions.

Examples of this procedure are philosophers Dagfinn Føllesdal (1979) and Chrysostomos Mantzavinos (2005) whom both are rejecting hermeneutics as method. Mind you, hermeneutics is not a "scientific" method, according to Føllesdal and Mantzavinos, in their view it is simply a futile and unnecessary exercise; there are the more precise methods applied by the natural sciences.[3] Their theory is devoid of any mention of the genealogy of interpretation within the perceiving and cogitating subject—a cognition that is ***embodied*** and ***enacted***,[4] to which we will return.

Føllesdal and Mantzavinos both find the hypothetical-deductive method to be the better choice. This is, in their view, a real rational alternative to any individual's interpretation. Interpretations are fraught with the unknowns of the human mind that cannot be anticipated or calculated—or at all ***known***, and should therefore, according to both Føllesdal and Mantzavinos, be neglected. However, Føllesdal finds that the hypothetical-deductive method alone is not enough to permanently close the door on the unknowns of the individual. "Decision theory" and "game theory" are needed for any deliberation to be kept at a scientifically speaking satisfying representational level.

The critique presented by Føllesdal and Mantzavinos is a critique aimed particularly at Heidegger and Gadamer. Dagfinn Føllesdal proposes that the hermeneutic method equates to the hypothetico-deductive approach when applied to meaningful content, such as texts, artworks, and actions. Føllesdal's criticism is directed toward those hermeneutists who confine the hypothetico-deductive approach solely to the realm of natural sciences. Here Føllesdal refers to the Diltheyian separation of the humanities and the natural sciences, enhanced by Snow's *The Two Cultures*[5] to consist of, as Snow puts it, "literary intellectuals" on the one side and "scientists" on the other. Hermeneutists like Joseph Kockelmans, Robert Crease, Patrick Heelan, Joseph Rouse, Hubert Dreyfus, and Don Ihde have argued that hermeneutics, or interpretation, is just as characteristic of the

[3] I have earlier presented Føllesdal's and Mantzavinos arguments in Friis (2016a, 2016b).
[4] Friis (2016a).
[5] C. P. Snow (1963).

natural sciences as of any other human mode of understanding. Føllesdal turns this claim 180 degrees around.

Føllesdal's focus is exclusively on the method; he does not delve into the question of "understanding"—even though interpreting is a deliberation of "necessary" empirical consequences from hypothesis and premises. Føllesdal contends that the H-D method is perfectly suited for meaningful materials such as texts and artworks.[6] However, Føllesdal's primary concern lies in the rationality of the method rather than in grasping what happens in an event, a facet he minimizes, perhaps due to Gadamer's assertion that understanding itself is an event that occurs suddenly and eludes conscious, rational control and manipulation.

Føllesdal insists that the process of scientific "deliberation" (rather than hermeneutic interpretation) must be governed by logic. Føllesdal's concern revolves around the rationality of the method itself. His interest lies in transcending the "relativism" brought about by an individual's historical and traditional context, as these factors impact the interpretative outcome. Føllesdal's logic is rooted in decision theory, the hypothetical-deductive method, and game theory. He firmly maintains, "No satisfactory study of man can take place without game theory."[7]

Føllesdal rejects the idea that humans should be inherently irrational or that unknown or tacit unconscious psychological processes drive most of an individual's actions—it is not rationally useful to operate with unknown unknowns. According to Føllesdal, in decision theory human actions are rational actions with premises known to the decision theorist. Initially, a person assesses the possible alternatives in the given situation. Subsequently, they choose one alternative that, considering their values and beliefs about probabilities, maximizes their expected utility. This deductive or conclusive process occurs within the framework of decision theory analysis, a rational process devoid of unknowns. Here, the hermeneutical circle, described by Gadamer as the point where the division between subject and object dissolves during interpretation, is replaced by the gradual testing of test implications derived logically from the proposed hypothesis.[8] The problem with the hermeneutical circle is, from a natural scientific methodical point of view, that understanding already presupposes what an object

[6] Friis (2016a).
[7] Føllesdal (1979).
[8] Friis (2016a).

is—it already presupposes a definition (understanding) of that object. Science does not allow circular explanations of this kind.[9]

In essence, a rational interpretation of a person's actions necessitates that "a rational individual, placed amidst multiple alternatives to be realized at time *t*, makes their choice based on transitive preferences at *t*."[10] To draw conclusions, we anticipate a person's preferences to remain consistent over time. Føllesdal clarifies that this does not mean a person cannot change their preferences; rather, it is crucial to comprehend why such changes occur. Rationality dictates that preferences are not solely guided by immediate desires but also by considerations of one's own future.[11]

Føllesdal believes it is time to replace hermeneutics with something much more quantifiably robust. In this view, understanding and interpretation become rational processes devoid of unknowns. The hermeneutical circle is supplanted by systematically testing empirical consequences derived logically from the proposed hypothesis. The interpreter's primary concern is to uphold a rational procedure, guided by principles such as temporal consistency, future-oriented considerations, and interactions between agents. In essence, a rational interpretation of a person's actions necessitates that a rational individual, when faced with various alternatives at a given time, makes choices based on transitive preferences.

Lastly, Føllesdal introduces the concept of interactions between individuals, suggesting that game theory is essential to comprehend these dynamics. He contends that a comprehensive study of human behavior requires the inclusion of game theory.

We must inquire why Føllesdal's methodological approach is not considered a more precise analytical tool compared to individual interpretation. The material under analysis possesses inherent meaning, and, in Gadamer's terms, understanding is the product of interpretation. How can one achieve understanding through the application of Føllesdal's intricate methods? Even with the implementation of H-D M, interpretation remains indispensable; meaning inevitably emerges through it. Without understanding, Føllesdal's methods are rendered meaningless. The circular nature of pre-understanding is concealed behind hypotheses, premises, and empirical consequences. In any deliberative process, the interpretive

[9] Érdi and Tsuda (2002: 310).
[10] Føllesdal (1979).
[11] Friis (2016a).

cognition that generates meaning is always tacitly or pre-consciously present.

Interpretation has its limitations. In radiology, there exists a phase where the perception and interpretation of an image occur pre-consciously, devoid of intentional influence or volition. The emergence of meaning that forms in consciousness is beyond individual control. These limitations stem from an individual's prehistory or horizon, his or her skill set, cultural identity, social demography, and psychology, collectively constituting their situatedness. Moreover, biological constraints limit our cognitive reach. The solution is not to forsake hermeneutics but to extend its scope to encompass contemporary fields such as science, technology, AI-human interaction, and medicine.

In radiology, psychological limitations are many. Consider a real case where the psychology of two radiologists almost led to an underdiagnosis. A girl presented at the ER with wrist pain following a bicycle accident, prompting an X-ray of her wrist and forearm. The next morning, two radiologists examined the image but failed to detect the wrist fracture suggested by the ER. A third radiologist, entering the room and unaware of the prior focus, immediately identified a fracture higher up the forearm. This phenomenon, common in radiology, demonstrates that even experienced radiologists can be misled.

This scenario underscores issues related to the reliability of interpretation. The interpreter's psychology is both an asset and a liability, influenced by the pre-understanding that shapes their focus. Additional perspectives may help, but not if all interpreters share the same erroneous assumptions, leading to identical predispositions. The third radiologist's lack of the pre-understanding provided by ER expanded the search area, resulting in a correct diagnosis. This either illustrates that the initial stages of interpretation, guided by a specified pre-understanding, can reveal a pattern indicating a fracture or miss it entirely if the pattern is absent in the area suggested by ER. Subsequent stages should reflect on the lack of findings despite patient symptoms, prompting a broader analysis.

Hermeneutics is as effective in radiology as the radiologist's willingness to apply it properly. Yet, scientific inquiry is fraught with socially and culturally formed biases that no procedure can fully eliminate. As Gadamer states, "When a naïve faith in scientific method denies the existence of effective history, there can be an actual deformation of knowledge."[12] In

[12] Gadamer (2007: 301).

the sciences, our perspective for interpretation is shaped by our experiences and education. However, in fields like hermeneutics, particularly as applied in disciplines such as cardiology and radiology, there is an emphasis on embracing interpretative diversity and inclusive practices. Unlike approaches that aim to eliminate subjectivity, hermeneutics values the varied perspectives that different individuals bring to the interpretive process. Thus, inviting others to review our interpretations, as done in the humanities, allows for the involvement of different horizons. This engagement goes beyond a mere fusion of horizons, recognizing that our perspective differs from others and thereby overcoming particularities on both sides.[13] We expand through involving.

However, Føllesdal seems to have altered his position somewhat. In 1993 he and a group of hermeneutists met in Hungary, discussing the role and scope of hermeneutics. Føllesdal and Karl Otto Apel defended the position that *hermeneutics is for the humanities only*. On the other side were Robert Crease, Lazslo Ropolyi, Don Ihde, and others, arguing that hermeneutics should "be expanded to the nonhuman, the inorganic, and artifactual as well, but in a unique P-H (philosophical-hermeneutics) sense."[14]

A more recent voice in this discussion is Mantzavinos. In his work, Mantzavinos concentrates on the viewpoints of Dilthey, Heidegger, and Gadamer—particularly these three figures as pivotal influences on the surge of hermeneutics in the French- and English-speaking spheres. Like the early Føllesdal, Mantzavinos dismisses hermeneutics as a viable avenue for pursuing understanding. He also aligns with Føllesdal in advocating for the hypothetico-deductive approach. Although there are nuanced differences between Mantzavinos and Føllesdal, where Føllesdal acknowledges the value of both traditions, Mantzavinos advocates for a complete abandonment of hermeneutics as a worthwhile philosophical pursuit. In Mantzavinos's words:

> …the philosophical hermeneutics so predominant in the German-speaking world is afflicted with many grave deficiencies and contains little of use. In particular, the hermeneutic views of Heidegger and Gadamer offer no satisfactory solutions – either to the problem of text interpretation or to any other problems. Heidegger's philosophy, developed with impressively

[13] Ibid.: 305.
[14] Ihde (1998: 40).

> formulated, excessive claims, in principle offers nothing more than a series of trivialities that are practically devoid of information. Gadamer's conception, although more closely connected with the traditional hermeneutic questions, possesses a very low problem-solving capacity: The transcendental vocabulary and the claim to the universality of the approach are not sufficiently substantiated, and they offer neither a correct analysis of the process of understanding nor a useful methodological guide for interpretative praxis.[15]

These words highlight perfectly the controversy between the two traditions. The analytical tradition is post-positivistic, and in the writing of the earlier Føllesdal and Mantzavinos upholding a notion of a unity of scientific method across scientific disciplines. We should not attribute this characteristic to all analytic philosophers; analytic philosophy is as diverse as continental philosophies. Bas C. Van Fraassen is one embracing methodological diversity, stating, "(...) we do have concrete, recognizable instances of other cognitively significant approaches (...) When the other person appears as a mystery, when the encounter is apprehended as contact with another world only overlapping one's own, when experience no longer fits everyday pigeonholes, that is the dawn of true understanding. When the artwork (...) reveals its own inner necessity, and one sees how it goes beyond past art by breaking yet fulfilling past demands, we break through to aesthetic cognition."[16]

The continental tradition is anti-positivistic in the sense that it is unthinkable that the humanistic sciences should apply the quantitative methods of the natural sciences. The continental side has embraced the notion of "rationality" within the seemingly irrational.[17] Hence, philosophical hermeneutics have an easy conscience rejecting the idea that understanding can be achieved through the serenity of logical reasoning alone.

Nevertheless, one does not simply eliminate the subjective influence just because one changes the method. We really do not know how to eliminate motives and intentions that are inherent in the research context or subdue intruding interests or needs or emotions resulting from the

[15] Mantzavinos (2005: 155).

[16] Van Fraassen (2002: 170–172).

[17] "Irrational" is here understood as the cognitive activity *preceding* conscious awareness, in other words, an action-determining-activity *transcending* or going *beyond* any willed or rational control.

activity taking place—these bodily embedded psychic elements are invisibly staining the fabric of almost any decision we make. So, are we then simply *pretending* that the opaqueness of subjectivity is a non-existent feature of human scientific thought and practice? Are we intentionally able to set aside emotions and whims surfacing unexpectedly with our activities? Is the idea that a scientist only applies pure logical thought, carefully calculating real values, in reality no more than wishful thinking? Norwood R. Hanson was skeptical of overemphasizing the value of hypothetical-deductive reasoning—of which—he thinks—there is something wrong: "If it (Hypothetical-Deductive Method) were construed as an account of physical practice it would be misleading. Physicists do not start from hypotheses; they start from data. By the time a law has been fixed into an H-D system, really original physical thinking is over."[18]

We need to retrace our steps and go back to the issue of *understanding* that is characteristic of hermeneutics. Gadamer published his *Wahrheit und Methode* in 1960—on the occasion of his 60th birthday, and his first publication in twenty-nine years.[19] Here he states that understanding generates truths determined by its object and that various sciences, like the humanities, generate different kinds of truths, its own truths, is a central theme of Gadamer's work. Gadamer unfortunately, like Heidegger, ends his hermeneutics with setting the goal of understanding as *self-understanding*. In the decades following Gadamer's groundbreaking work hermeneutics have taken on many forms and functions, the important development for this volume is the fact that interpretation as a scientific activity is not only aiming at understanding itself but also the other, that which is not self. However, there is no need to become suspicious, we are not mimicking the exact sciences when we apply hermeneutics on the "other." The exact sciences also interpret, as is a well-established fact, however, this does not make the exact sciences less accurate. We engage the world as individuals as we also do in the sciences, but differently.

Hermeneutical acts does not begin from a *tabula rasa*—or as Grondin writes—perception is never entirely direct, "in every understanding there is a fusion of the present and the past (…) there is never a complete rupture between the two."[20] In the understanding, generative cognitive process, whose origin is temporarily prior to analytical thinking and typical of

[18] Hanson (1958: 70).
[19] Grondin (2022: 2).
[20] Ibid.: 5.

deductive reasoning, feeds on perceptions of the world as well as of prior understandings. This is how any new understanding is generated. In cognition the present is a continuation of the past. Culturally speaking we are all heirs to open-ended cultural traditions, to series of previous related thinking making up a tradition, which, through education, constitutes our own knowledge base. There are countless many narratives, ours is but one narrative. This answers—at least in part—the question *why* we perceive what we do out of the enormous mass of stimuli in the world. Birds can for instance see further and with more colors than we do. Our physical being also sets limits to what and how we can know or understand. Cognitively speaking, interpretative activity tacitly precedes conscious willed activities—generating the *impetus* for such activities the same way future activities will take its beginning from the understanding of the present.

Martin Buber[21] wrote about the same contradistinction between the two mental modes of "relating to the world." Buber describes calculative and detached "relating to" as an "I-It" attitude toward an object that is separate in itself, which we experience and perhaps choose to make use of. And, as Buber says, we are categorizing the object by its type; that is, studying it; exercising the ability to look at something from a different perspective; and, in fact, *interpreting* the experience of the object in functional and mathematical terms.[22]

The other mode of thinking is, in its imagined separateness from calculative thought, very different and Heidegger has termed this activity "meditative," and which is *engaged perceptual understanding* or, simply, engaging with the world. Then something happens with us—within us. There are no givens to the senses but an environment of *significances.* In Buber's terms, this means that we perceive the object in front of us with color, detail, fragrances, and movements that *mean* something to us. We are interpreting the object into what we see. To Buber this is the attitude of the "I" toward "Thou," a relation in which the other is not separated from the person that sees or perceives the person opposite—and if we bring Levinas's ethics into this image, I am confronted with the face of the other with all that it expresses without a separation between I and that which the face expresses. The face is not just an object I perceive; it is a presence that commands me, a vulnerability that calls me to respond. This

[21] Buber (1970).
[22] Ibid.

moment disrupts the primacy of the self and reveals a meaning that is prior to any constructed identity. Interestingly, *interpretation* is in Buber's mind the fundamental cognitive process of both calculative and meditative thinking even though the two modes, torn apart and imagined as separate activities, one conscious and the other sub- or pre-conscious, stand in theoretical contradistinction to each other. It is, however, in the meditative mode of thought that interpretation is a "relating to" something else with an understanding that is without objectifying. This is embodied interpretation, an enactment that systematized has become the method of hermeneutics.

To elaborate, there is a most interesting distinction between Heidegger and Gadamer. This difference between the two is referred to as the third phase of hermeneutics.[23] For Heidegger hermeneutics is not about interpreting written works. Our aim is never to understand because understanding is something *we already do*. Understanding is instead the precondition of interpretation. In looking at images understanding is disclosed, that is, an understanding that appears from recontextualized memory. This is, according to Heidegger, altogether distinct from any theory of representation or propositional knowledge.[24]

Nevertheless, in both Heidegger and Gadamer's hermeneutical approaches, which is termed "philosophical hermeneutics," there is no activity, entity, or institution that is not interpretable. As sociologist Barry Sandywell points out, *all thought has a hermeneutic dimension*. One important insight concerning understanding, according to Sandywell, is that all understanding already presupposes understanding that has been mediated by dialogical forms. All is therefore ontologically informed, and all access to reality is possible only as mediated existence. All human activity and practices can therefore be interpreted and understood in their social and historical contexts.[25]

Interpretation is the "core activity" of any human's everyday life. From the perspective of logic and rational, systematic thinking, interpretation can be methodical and rule based. However, meanings, the *significances*, at the center of the analysis, are established *prior* to its more logical deliberations during the analysis. Meaning-making occurs at a *pre-conscious* and neuronal level, an activity phase where the ongoing neural and chemical

[23] Davey (1999).
[24] Friis (2020b).
[25] Sandywell (2011: 324–325).

processes constitute what we term an *act—which is bodily-cognitive interactivity generating actions, perceptions, and conscious thought*, an ongoing activity that cannot be grasped in its activity. Tarkovsky explains the interpretative modes of perceptually based thinking beautifully—not as a way of calculation but as a state of *meditative* mind, of being open to being, of awaiting:

> ... the empirical process of intellectual cognition cannot explain how an artistic Image comes into being – unique, indivisible, created and existing on some plane other than that of the intellect. Here it is a question of agreeing on terminology. In science, at the moment of discovery, logic is replaced by intuition. In art, as in religion, intuition is tantamount to conviction, to faith. It is a state of mind, not a way of thinking. Science is empirical, whereas the conception of images is governed by the dynamic of Revelation. It's a question of sudden flashes of illumination – like scales falling from the eyes; not in relation to the parts, however, but to the whole, to the infinite, to what does not fit into conscious thought.[26]

Embodied hermeneutics, the innate ability to pre-consciously interpret—to grasp, goes into hiding, or as Heidegger terms it, becomes *veiled*, as soon as discursive, calculative, and discriminative thought, the "I-It" relation to objects, is allowed to take over.

The objectifying cognitive modus hailed as rational and as the very essence of humanity has grown out of an increasing and insatiable appetite for images and visual forms of knowledge. Modern science is fixated upon the visual.[27] Modernity's relation to lived experience, to sense experience, is complex and conflicting. Traditionally, science both needs it and excludes it. Visual experience appears to be natural, yet, it is a cultural variable of socio-historical formation.[28] This is the case with radiology. It is technologically fixated upon the visual, yet, in spite of measurements and ways of manipulating the modalities and PACS to render the images as clear, sharp, and detailed as possibly can be, the visual experience and the reporting of findings can be *imprecise*.

It is *embodied* hermeneutics, the visual experience of the radiologists, their gaze, and interpretations—the issue of understanding what is at stake that this book sets out to scrutinize in order to shed light on the medical

[26] Tarkovsky (1989: 40–41).
[27] Sandywell (2011: 3).
[28] Ibid.: 10.

practice of interpreting. However, not what it ideally should be but what interpretation is when performed—whatever result, and with diagnostic certainty and accuracy as the backdrop to which it is "measured" up and against.

Human cognition is fundamentally a meaning-making activity, arising from the integrated use of the body, sensory organs, brain, and mind to interpret and act upon affordances in the environment.[29] This process is an embodied, engaged activity, shaped and cultivated by customs, practices, and institutional contexts. As Thompson argues, the concept of "enaction" emphasizes the action-oriented nature of cognition—highlighting that humans possess a capacity for interaction that is not always the result of rational deliberation or explicit representation.[30] Rather, these cognitive abilities develop and expand through dynamic, ongoing interactions with others and through immersion in sociocultural environments.[31]

The embodied nature of human cognition is further illuminated by Varela, Thompson, and Rosch, who assert that "cognition depends upon the kinds of experience that come from having a body with various sensorimotor capacities… and these sensorimotor capacities are themselves embedded in a more encompassing biological, psychological, and cultural context."[32] Humans, as "body-brains," construct memory through bodily actions, drawing upon experiences in a pre-reflective manner. Understanding, in this view, is not simply a cognitive endpoint but a prerequisite for engagement; it often evolves through the very process of interaction, resulting in a "widened" or "deeper" understanding that may differ significantly from one's initial perspective. Acting on assumed knowledge, grounded in experience, facilitates these engagements. In contexts such as interpreting radiological images or interacting with objects and people, our embeddedness within a specific context implicitly guides our responses. This tacit guidance is an example of memory recontextualization—an automatic form of engagement that occurs prior to conscious, rational deliberation.

This phenomenon can also be observed in everyday actions, such as working at a desk, where tasks like writing, drinking coffee, or searching for a pen often occur simultaneously and with little conscious effort. Even

[29] Hutto (2013: 2).
[30] Thompson (2007: 12).
[31] Hutto (2013: 4).
[32] Varela et al. (1993: 172–3).

in such seemingly automatic actions, cognitive processes unfold to help locate objects or coordinate movements, although these efforts largely go unnoticed. Over time and with repetition, these engagements are stored as "skill memory" or "habit memory," embedding learned actions within one's repertoire of intuitive behaviors.[33]

Humans also attune to perceptible environments and adapt to cultural and social contexts. In sociocultural settings, individuals acquire fluency in interacting with tangible objects, particularly through sustained training and practice, as with operating complex technologies. The increasing sophistication of technological tools, such as advanced MRI or modern X-ray imaging, has exposed certain cognitive limitations in humans. As these technologies evolve, they often bring unforeseen challenges, demonstrating the adaptive, yet imperfect, nature of human cognition in response to complex, technologically mediated environments.

Despite the existence of extensive written procedures and guidelines designed to support radiologists, these resources have done little to mitigate cognitive biases. In fact, reliance on these documents may even exacerbate the issue. Libraries are stocked with manuals and performance guides that remain largely unused in daily practice, as radiologists rely on peer mentorship for learning and conform to the interpretative methods modeled by professors or supervisors. Over time, this informal training system solidifies into local norms that govern interpretative practices. However, these norms are byproducts of individual interactions and thus cannot fully address the inherent cognitive biases that persist in radiological interpretation. One contributing factor to these biases is the apparent cognitive strain involved in the complex human-technology interactions within radiology.

Theoretically, it is challenging—if not impossible—to achieve both intersubjectivity and reliability in image readings. Intersubjectivity refers to the consistency between different radiologists' interpretations of the same image, while reliability refers to the consistency in a single radiologist's interpretations of the same image over time. However, radiological practice frequently exhibits both inter-observer and intra-observer variability: radiologists in different institutions or even within the same hospital may interpret the same image differently. Similarly, a single radiologist might interpret the same image differently over time. This variability

[33] Casey (1984).

highlights the difficulty of achieving interpretative consistency, as contextual and individual differences continually shape perception.

Embodied cognition, particularly within the framework of hermeneutics, can provide insight into these challenges. As radiologists interact with Picture Archiving and Communication Systems (PACS), observe images, and identify patterns, they draw upon embodied, often subconscious knowledge. This process involves not only recognizing shapes or anomalies but also mobilizing experience-driven, pre-conscious understanding. Humans engage with objects and technologies by perceiving, interpreting, and physically interacting with them, drawing upon stored knowledge and experiences gained through training. Yet human interpretative abilities are inherently limited when compared to the precision offered by scientific visualizations. Technologies are developed precisely to mitigate these perceptual limitations, and while they extend human capability, they also introduce cognitive constraints, as users must adapt to technological interfaces and workflows.

Beyond perceptual limitations, human cognition is context-bound; we cannot fully recall or process every past experience or thought, as context continuously reshapes memory and perception. Our conscious awareness is, therefore, often limited to fragments of previous experiences. This dynamic interplay of partial recollections, which we continuously project onto present situations, means that we navigate reality through rehearsed "bits and pieces" of lived experience. Certain aspects of the world—the microscopic, the distant, and the imperceptible—are inherently beyond our grasp. This partial and often approximate cognition shapes not only our beliefs and emotions but also the tacit interpretative frameworks we apply to make sense of the world. Consequently, incomplete knowledge fundamentally underpins human experience.

In the context of radiology, the problem of incomplete knowledge becomes even more pronounced due to the precision demanded by the field. While the accuracy of scientific visualizations plays a minimal role in day-to-day life, in radiology, the pursuit of fine-grained accuracy is paramount. However, humans tend to rely on science, authority, and sensory experience to construct a reliable understanding of the world, often combining notions of absolute and relative truths. Demands for extreme accuracy, such as those in mathematics or technology, push the boundaries of human perceptual capabilities, further highlighting the limitations of embodied cognition.

The analysis of an image appears to be a straightforward and methodical affair—and it is up to a certain point. The rational procedure of systematics quickly becomes fuzzy. The radiologist performs rationally. He follows the procedures and hence experiences his performance as rational and systematical. Identification or lack hereof comes tacitly and suddenly according to expectation or contrary to it. He describes orally what he has seen and identified. In other words, the radiologist performs both modes of thought simultaneously. He is methodical and performs a systematical "go-through" of the images, his approach is rule based, yet, his recognition and identification of the clues of the pattern hidden in the greyscale of distributed pixels is intuitive and tacit—a performance based on experience, embodied skills, and memory of learned patterns. The radiologist experiences no tension what so ever between the modes of a cognitive performance simultaneously enacting the results. However, during the radiologist's gaze at the screen, during that very moment, something significant happens that is beyond her or his cool, beyond rational control and with consequences for the report—and ultimately for the patient.

The aim of this book is to provide an interpretive exploration of image perception and embodied hermeneutic cognition, drawing on a range of theoretical perspectives. In doing so, I will also examine the challenges posed by cognitive variability in radiological interpretation, analyzing both the strategies that have been ineffective and the practices radiologists employ to enhance diagnostic accuracy. By synthesizing these insights, the book seeks not only to illuminate the complexities of radiological cognition but also to contribute to ongoing efforts to improve interpretational precision in medical imaging diagnostics.

Hermeneutics, in this context, emerges as a process of amplification—an enhancement of human interpretative faculties that engages with a world extending beyond the realm of human-created artifacts. This book, therefore, illustrates hermeneutics' versatility and significance across disciplines by equipping the human mind, regardless of its specific area of inquiry, with the capacity to generate understanding and meaning. In contemporary philosophy, there is a revitalized focus on the interplay between interpretation and scientific practice, particularly regarding how observation in the sciences is inherently shaped by hermeneutical dimensions.[34] This perspective challenges the traditional dichotomy between the human-

[34] See Mejer (2022)—*Updating the Interpretative Turn*; and also Canaris (2017)—chapter 1: Major Trends in Modern Hermeneutics.

ities and the sciences, emphasizing the interpretative frameworks underlying fields often associated with objectivity. As a result, hermeneutics is expanding its traditional boundaries to address issues central to scientific investigation, demonstrating its growing interdisciplinary relevance within the scientific practices it engages. Therefore, a hermeneutics of scientific practice—especially one concerned with the hermeneutics of technology as an extension of human experience beyond direct perception—must engage with the *origins* of perceptual or phenomenal gestalts, which form the basis for understanding.

Embodied hermeneutical cognition does not attempt to explain rational, conscious deliberation explicitly. Rather, it offers a plausible framework for understanding human cognition as it emerges from neural and bodily interactions with the world. It explores how humans adapt, attune, and relate to others, to their environments, to technology, and to their own lived experiences. This perspective does not undermine the importance of rationality or humanity's capacity for rational thought; instead, it provides a foundational context for understanding how pre-reflective, embodied processes shape cognition and meaning-making within the constraints of human experience.

CHAPTER 2

# Disruptive Cognitions, and Interpretation in Radiological Image Analysis

Heidegger asserts that understanding is not something we acquire upon entering a situation; rather, it is an activity we are already engaged in.[1] This insight applies directly to radiology, where image interpretation involves much more than merely observing images on a PACS. The interpretative process is shaped by the radiologist's accumulated expertise, medical training, and the specific preparation for the case at hand. More experienced radiologists, in fact, often demonstrate greater precision in their diagnoses, sometimes bypassing established systematic procedures because they can intuitively recognize key features from their first glance at an image. This instinctive recognition reflects their advanced skill, but it is also important to acknowledge that interpretation remains the central cognitive task for the medical professional. Medicine, and radiology in particular, inherently relies on this interpretative knowhow of its practitioners. However, the reliance on human interpretation also introduces the potential for error, which drives researchers and engineers to explore alternatives. Error in radiology is in fact defined by inaccuracies in the interpretation of images resulting in diagnostic mistakes. Errors happen due to human, cognitive, technical, or systemic factors.[2] Efforts to minimize the subjective variability of human cognition by turning to quantifiable, ostensibly objective methods drawn from the natural sciences are often undertaken

[1] Heidegger (1992a, § 32).
[2] Bruno et al. (2015).



J. K. B. O. Friis, *Hermeneutics at the Intersection of Medical Technology*, https://doi.org/10.1007/978-3-031-88755-0_2

in an attempt to reduce the risk of misdiagnosis, standardize assessment, and enhance clinical reliability.

Radiology is a medical specialty that employs imaging technologies such as X-rays, MRI, CT scans, and others to produce visual representations of the internal structures of the human body. These technologies create detailed, diagrammatic depictions of organs, tissues, and systems—such as muscles, the brain, arteries, lungs, and the skeleton—that can be interpreted by a trained specialist. The visual content generated by these imaging techniques can be understood as a *gestalt*, a concept that refers to the holistic pattern recognized and identified by the radiologist. This *gestalt* emerges in the moment when the radiologist perceives and becomes consciously aware of a meaningful pattern in the image. Importantly, the *gestalt* exists only in the mind of the observer, as it is the cognitive recognition of the underlying structure or anomaly that makes the image clinically meaningful. Thus, radiological interpretation is not merely the passive reception of visual data, but an active cognitive process of pattern recognition and meaning-making.

Radiologists are acutely aware that cognitive biases pose significant challenges to the interpretation of radiological images. As Richard L. Gregory notes in his essay "Putting Illusions in Their Place," "By revealing unconscious knowledge, assumptions, and operating rules of vision, illusions reveal hidden depths of our minds." This insight highlights the profound role of unconscious processes in visual perception, which can lead to interpretive errors. In response to this issue, researchers have sought solutions, including the introduction of standardized protocols as a form of "cognitive fix." Others have proposed quantitative approaches aimed at circumventing the unpredictability arising from what Gregory describes as the "hidden depths of the mind."[3]

Cognitive biases emerge during visual perception when irrelevant memories interfere with the accurate reading of an image. The formation of *gestalts* relies on memory traces, which are shaped by training, education, and experience. These long-term memory traces, often referred to as "skills," are not static but dynamically recontextualized based on the task or context at hand. During visual analysis, these recontextualized memories interact with neural processes, such as transitive inferences,[4] influencing

[3] Gregory (1994: 260).

[4] Transitive inference happens during neural processing and is the ability to infer a relation between image patterns, situations, people, objects, events, that have not been presented together based on previous memory.

the interpretation of the image.[5] If the radiologist's memory lacks an accurate trace corresponding to the pattern presented in the image, an incorrect *gestalt* may form, leading to misidentification. In such cases, a pattern different from the one depicted in the image is perceived in the radiologist's mind.

Another crucial factor in radiological cognition is the neurological and physiological structure of the brain, which underlies the act of perception. Human visual perception is governed by what can be termed the "predetermined possibilities and limitations" of our sensory and neural systems, which define the range of detail we can perceive. This perceptual framework also sets limits on what can be processed at the neural level. A significant temporal delay exists between the moment light from an object reaches the retina and when the corresponding *gestalt* enters conscious awareness—a delay of 0.3 to 0.5 seconds. During this time, the brain compensates by drawing on recontextualized memories or experiences to anticipate patterns, generating *gestalts* before full conscious awareness. This process enables individuals to react swiftly to real-world situations, such as instinctively jumping out of the way of a speeding car.

However, despite these adaptive mechanisms, cognitive biases can still lead to errors in radiological interpretation. Before exploring the statistical frequency of such errors, it is important to recognize that these biases arise not only from the intricacies of memory and perception but also from the broader limitations inherent in the human brain's design.

## 2.1 The Bias We Live with

In radiology, cognitive bias arises when patterns in diagnostic images become subtly distorted during interpretation by the observer. A major contributor to this distortion is observer variability. Radiological errors are generally classified into three types: perceptual errors, cognitive (or interpretative) errors, and systemic errors.

Lev, Rhea, and Bramson, in *The Lancet*,[6] reported that observer variability has serious clinical implications. Their necropsy studies revealed

[5] Heckers et al. (2004); Frank et al. (2006).

[6] Lev, Rhea, and Bramson, *The Lancet* (1999). Although these data are "old," it is unlikely that the fundamental mechanisms or timing of cognitive biases have changed significantly since 1999. These data serve to illustrate a persistent reality: cognitive biases are deeply ingrained and operate largely outside our conscious awareness, subtly influencing both our perceptions and actions. Such biases continue to shape our interpretations of information and our decision-making processes, often without us realizing their impact.

that up to 40% of patients may not receive a timely radiological diagnosis to prevent fatality, with diagnostic errors occurring in approximately 30% of cases. While these figures are somewhat dated and may vary across countries and practices, recent research by Itri and Patel supports similar estimates, finding diagnostic errors to range between 20% and 40%.[7] Perceptual errors, by the way, account for approximately 60–80% of radiological errors. Systemic errors are technological, like poor image quality, procedure may fail or findings are misreported or uncommunicated.[8]

Much of the research in radiology today centers on identifying and characterizing the cognitive and perceptual challenges intrinsic to diagnostic interpretation. Scholars seek to uncover cognitive "trap doors" and to develop systematic approaches aimed at reducing interpretive variability and perceptual inaccuracy, often advocating for standardized protocols to ensure consistency across practitioners. While these methodological solutions assume a rational, systematic cognitive process, we have to contend that such a view is only part of the picture. Human cognition is not inherently designed to deliver "true-to-nature" perceptions; even with a systematic approach, radiologists are constrained by their inherent perceptual limits. As imaging technology advances and produces ever more detailed data, the volume of information in each image increasingly exceeds the radiologist's ability to process it exhaustively.[9]

Ultimately, the core issues in radiology are rooted in the perceptual and cognitive limitations of the human observer. Radiologists engage with mediated representations, and it is precisely during this interpretive interaction that perceptual and cognitive challenges arise. Interestingly, while these limitations pose obstacles in a clinical setting, they do not impede general human cognition in daily life, where judgments and biases occur regularly without noticeable negative effects. In our everyday environments, biases and perceptual illusions seldom lead to serious consequences, and we are often unaware of their influence.

This is not to undervalue human cognitive capacities. As embodied beings, we navigate life through perceptual judgments and cognitive processes that align with experiential norms. However, as Louis-Ferdinand Céline insightfully noted, "experience is a dim lamp, which only lights the one who bears it." All experiences are ultimately personal, informed by

[7] Itri and Patel (2018).
[8] Donald and Barnard (2012).
[9] Friis (2015b, 2020a, 2021).

individual memory traces that shape our perceptions, evaluations, and interpretations. For the radiologist, these cognitive-perceptual capacities can become limitations when they prevent the detection of subtle abnormalities, such as a small cancerous nodule concealed within normal tissue.

Kundel[10] asserts that perceptual image analysis is the cornerstone of a radiologist's role. Radiologists typically regard their interpretations as objective reflections of the image content, rarely considering the perceptual process—except in cases where errors arise. However, the interpretive process is far from straightforward, as it is influenced by a range of cognitive and environmental factors, many of which operate outside conscious awareness. Among these challenges, Krupinski[11] highlights anatomical overlap, which can obscure lesions and hinder detection. While advanced imaging technologies like PET and CT scans address this issue by generating cross-sectional views that reduce overlap, the burden remains on radiologists to discern subtle yet critical details within each image slice to ensure accurate lesion identification.[12]

It should be mentioned that to reduce radiological errors, there are strategies targeting human, cognitive, as well as systemic factors. One key approach is double reading, where two radiologists independently review high-risk cases.[13] Artificial intelligence (AI)-driven computer-aided detection (CAD) systems further enhance diagnostic accuracy by identifying anomalies and supporting radiologists in image interpretation.[14] Structured reporting, using standardized templates, is used to improve clarity and reduce ambiguity in radiological reports, while regular peer reviews are understood to foster quality assurance by allowing radiologists to learn from each other's interpretations.

Cognitive and training interventions also play a crucial role in error reduction. Bias-awareness training is thought to help radiologists recognize and mitigate cognitive biases that could affect decision-making. Simulation-based learning provides a controlled environment for practicing complex cases, improving diagnostic accuracy.[15] Additionally, experience-based calibration, in which radiologists receive feedback on past errors, helps refine their skills over time.

[10] Kundel (2006: 402).
[11] Krupinski (2011: 393).
[12] Ibid.
[13] Kundel et al. (2007).
[14] Giger (2018).
[15] Brady (2017).

Systemic solutions address workflow inefficiencies and communication barriers. Checklists and decision-support tools for high-risk conditions help prevent diagnostic oversight.[16] Standardized protocols improve communication between radiologists and referring physicians, reducing the likelihood of misinterpretation.

While technological advancements provide tools and procedures to enhance diagnostic precision and reduce errors, the interpretive act ultimately relies on the radiologist's perceptual and cognitive faculties. Whether analyzing two- or three-dimensional images, the radiologist remains tasked with navigating both the limitations of human perception and the complex, layered information within each scan.

More than 60% of diagnostic errors in radiology are perceptual in nature,[17] and these perceptual biases often stem from cultural learning and conditioning. Experiences in formal education, mentorship, play, and specialized medical training collectively shape not only the expertise but also the perceptual habits and biases of radiologists. While rigorous training aims to refine radiologists' attention to anatomical structures and abnormal patterns, it simultaneously modifies and constrains innate perceptual tendencies. Kundel[18] suggests that interpretive errors are introduced not only by clinical history but also by the radiologist's personal experiences and biases. A striking cognitive function illustrating this is the phenomenon of perceptual "fill-in"—the tendency for the mind to complete or anticipate missing information. This process, termed "unconscious inference" by Helmholtz,[19] highlights how perceptual conclusions are often constructed by the mind even before conscious awareness.

Helmholtz's work sets a constructivist framework within cognitive science, underscoring that perception is not a simple mirror of reality but rather a product of neural and chemical interactions within the body-brain system. What appears as a direct perception is, in fact, shaped by dynamic internal processes. The conscious self, then, is an adaptive construct, a practical self-formation that serves functional purposes rather than representing a direct interface with unmediated reality. As Thomas Merton[20] observed, whenever we distinguish, categorize, or judge, we impose layers

[16] Graber et al. (2014).
[17] Krupinski et al. (1998: 611–612).
[18] Kundel (1990).
[19] Friis (2011a, 2011b).
[20] Merton (1968: 7).

of interpretation on our perceptions. This notion echoes Heidegger's idea of an original "openness to Being"—the state of mind unburdened by social conditioning, historical influences, and specialized training. Within this innate openness lies a kind of "ontological generosity" or receptivity to reality that becomes shaped by our cultural framework.[21]

Poet Louise Glück captures this phenomenon poignantly, noting, "We look at the world once, in childhood. The rest is memory."[22] This line suggests that our initial, unmediated experience of the world is subsequently filtered through layers of memory and cultural understanding. Western cultural history, with its emphasis on rational, dichotomizing thought, has further distanced individuals from direct experiences of Being. As William Barrett[23] suggests, our intellectual traditions have not only made us technological masters but also distanced us from both the essence of our existence and our own being.

Thus, cultural and theoretical structures are inextricable from perception. The objects we see, whether tangible or virtual, are imbued with meanings shaped by language, culture, and memory. These meanings are dynamically interwoven with our remembered experiences and, often, with what we only believe we remember, surfacing in consciousness as familiarity when we encounter perceptually recognizable forms.

Consequently, perception is rarely pure or unbiased; it is a confluence of cultural conditioning, individual memory, and socially constructed meanings, all of which influence the radiologist's interpretive lens.

Despite occasional misinterpretations, we generally grasp the essential meaning conveyed in radiological descriptions. However, the fact that we do not always achieve accuracy suggests that our interpretations are not necessarily direct correspondences of the images we interpret. Instead, our perceptual engagement involves continuous neural and bodily enactments—neural-bodily activities that constitute the conscious reality we experience. Researchers such as Nobel Laureate Daniel Kahneman and neuroscientist Francisco Varela propose that our brain-body-mind complex adapts to various situations by drawing upon memory. This adaptive process enables us to rapidly identify objects, comprehend complex patterns, or even appreciate the layered meanings in a line of poetry.

[21] Ibid.: 25.

[22] Louise Glück (1968). *Notes from Firstborn.*

[23] Barrett (1956: xiii).

In relation to perceptual "filling-in" and "background" phenomena, noteworthy differences emerge between the young Heidegger in *Being and Time* and his student Gadamer. However, describing these distinctions as differences should not imply an opposition between early Heideggerian philosophy and Gadamer's hermeneutic framework. As noted by Davie,[24] Gadamer's understanding of meaning is often conceptualized in a text-like form. Should this interpretation hold, it would suggest a limitation within hermeneutics, confining it to textual or linguistic expressions rather than encompassing broader interpretive experiences and understanding. As Sandywell[25] argues, Gadamer's move can be viewed as a departure from the rhetoric of reflective knowledge toward a form of reflexivity that rejects Cartesian notions of detached, representational cognition as a pathway to truth. This shift emphasizes the figural and hermeneutic dimensions of understanding.

Building upon this view, Levin[26] suggests that moving away from the Cartesian framework of representationalism reveals the fundamentally perceptual nature of human beings.[27] We are, in essence, open to being—a concept that highlights how our interpretive acts engage us bodily, emotionally, and perceptually. Levin elaborates that "the gift of vision is a challenge; it is also an opportunity,"[28] but only if we ground our perception in bodily feeling and affective engagement. Our openness to being is thus an openness to be "touched and moved by what we see," an idea that aligns vision with tactile and emotional resonance. Levin's perspective draws on Heidegger's philosophy and echoes Merleau-Ponty's assertion that thinking is an embodied act.

We must critically examine Cartesian assumptions of disembodied cognition. By doing so, we restore a model of cognition that integrates its embodied, tactile, and affective dimensions, acknowledging the intersubjective, corporeal nature of human understanding and interpretation. This view reaffirms that cognition is not merely representational but deeply grounded in the human experience of perceiving and responding to the world.[29]

[24] Davey (1999).
[25] Sandywell (1999).
[26] Levin (1999).
[27] See Heywood (1999: 205 and 214), for a similar view.
[28] Levin (1999).
[29] Ibid.

In Heidegger's *Being and Time*, understanding is not an endpoint or objective of inquiry; rather, it is the foundational condition of human existence.[30] Understanding is both an inherent capacity and a defining characteristic of human beings, shaping our interactions and our very being. This process aligns with what enactivist theorists describe as "self-individuation"—an ongoing, dynamic interplay among human understanding, the world, and the self. Our sense of identity and our perception of reality are grounded in what we understand, as noted by Davey. In this way, understanding is not only fundamental but also a precondition for interpretation.[31] As already mentioned, Heidegger writes in *Being and Time*, "interpretation (is not) the acquiring of information about what is understood: it is rather the working out of possibilities projected in understanding."[32]

Gadamer, while reorienting Heidegger's perspective, does not discard it; rather, he shifts the focus to the immediacy of experiential understanding.[33] For him, perception is shaped by the convergence of background elements—what Nietzsche might call "perspective" and Gadamer terms "horizon" or historicity.[34]

In this interpretive space, vision and dialogue reveal subtleties of association and meaning, with much of the interpretive insight arising from the unsaid elements implied by what is spoken—a point emphasized by Davey. This perspective echoes Wittgenstein's assertion that "we must think of perception as containing a thought-element, and, perhaps, that we must think of thinking as containing a perception-element,"[35] highlighting the subtle interplay between cognition and perception.

In the context of radiological image interpretation, significant limitations exist concerning the reliability of perceptual accuracy. These limitations relate directly to the radiologist's cultivated perceptual memory, a skill honed by extensive training and experience.[36] Sight, in this specialized context, is not a fixed capacity determined solely by biology; instead, it is a potential developed and refined through social, cultural, and individual efforts. Radiologists face both inherent limitations of perception and the unique challenges posed by specialized motor-sensory demands, which are

[30] Davey (1999: 5).
[31] Ibid.
[32] Heidegger (1992a: 189).
[33] Davey (1999: 5).
[34] Ibid.
[35] Ibid.: 12.
[36] Levin (1999: 189).

continually tested and expanded throughout their training and professional practice.

The cognitivist framework posits that humans possess internal mental representations of external objects, representations that correspond to the truth or falsity of these objects. This perspective, which grounds cognition in a representationalist ontology of truth, reflects a distinctly modern viewpoint that Martin Heidegger critiqued for obscuring a more profound understanding of truth. Heidegger argued that truth, in its essence, cannot be reduced to propositional truth, which confines it to mere objectification. In *The Question Concerning Technology*, Heidegger remarked on a time when *techné*—understood as the craft of bringing forth truth into radiant presence—was synonymous with art. He described this bringing forth as a form of revealing grounded in piety, a yielding to truth's essence and preservation: "there was a time when it was not technology alone that bore the name techné. Once that revealing that brings forth truth into the splendor of radiant appearing was also called techné … And art was simply called techné. It was a single, manifold revealing. It was pious … i.e., yielding to the holding-sway and the safekeeping of truth."[37]

This critique of representation aligns with recent research on cognitive processes, emphasizing that human perception and understanding are inherently interpretative and influenced by memory and prior experience. The concept of "filling-in" reflects how memory and perception merge to allow anticipatory responses before conscious awareness fully registers the external world. In cases of perceived danger, for example, the body-brain system enacts an immediate response—an "irrational reason," or an embodied conjecture that precedes conscious deliberation. The temporal dynamics of perception reveal that this interpretative enaction operates faster than the conscious "coming to awareness" of an event. Indeed, much of the body-brain-mind activity remains unconscious. In any action, an individual's response emerges from the integration of memory, sensory input, and emotional states, allowing potentialities to transform into tangible reactions. This enactment is informed by, and simultaneously reshapes, sensory perception, emotional responses, and prior experiences, reconfigured for the current situation. As perceptual psychologist Steven Yantis[38] suggests, these mechanisms underpin veridical experience, even amid sensory limitations and distortions.

[37] Quoted from Levin (1999: 194–5).

[38] Yantis (2001: 21); also Friis (2011a, 2011b: 365).

The interpretative phenomenon of "filling-in" also challenges a strictly cognitivist approach when considering how perception functions within a physical system like the brain. While one could frame filling-in as a mental computation that informs the perceiver—aligning with cognitivist language—this interpretation overlooks the decentralized, dynamic interplay across the body-mind system. In perception, the senses, body, and brain participate in an ongoing exchange, generating emerging patterns that, in the perceiver's consciousness, manifest as gestalts. The truth-value of these patterns is not static but rather shaped by the perceiver's experiential background, namely, their memory. For instance, a radiologist discerns patterns within an image before they fully take shape as conscious gestalts and before any conscious manipulation occurs. These enactments, or pre-conscious recognitions, are situationally dependent and influenced by the individual's contextual knowledge and prior experiences, underscoring the complex and multi-layered nature of perception as an embodied, interpretative, and temporally fluid process.

A temporal structure exists within the brain-body system, which may be termed a "temporal loop." This structure operates both neuronally and chemically, manifesting as a form of volume signaling within the brain.[39] Unlike traditional neural wiring, where signals are transmitted along defined pathways, this system involves neuromodulators that diffuse freely in cloud-like formations across the brain, independent of cellular and membrane barriers. The specific nature of these temporal patterns can vary based on the type of triggering event, resulting in different temporal structures that influence both thought processes and behavioral responses.[40] For instance, in high-stress situations, the body-brain system may release stress hormones that drive a rapid transmission of "information" across cellular boundaries. This mechanism would enable information to be acted upon more swiftly than conventional neural signaling pathways allow, potentially bypassing certain temporal limitations of synaptic transmission.

This process highlights the close interdependence between the body and visual perception. A recent study conducted by researchers at the Max Planck Institute demonstrates that the brain receives and processes signals concurrently from both the body and the external environment, suggesting that the body plays a regulatory role in shaping perceptual attention.

[39] Wheeler (2014); Husbands et al. (1998).
[40] Wheeler (2014: 379).

Specifically, the study reveals that the body can both limit and enhance perceptual attention at different times. Despite these continuous, body-driven modulations of perception, we generally remain consciously unaware of them. For instance, although we are seldom consciously attuned to our heartbeats, these rhythms exert a significant, though hidden, impact on visual perception.

The Max Planck researchers identified two primary ways in which heartbeat dynamics influence perception: (1) increased neural activity associated with the heartbeat immediately before sensory stimulation correlates with reduced somatosensory detection, and (2) the timing of sensory stimuli within the cardiac cycle affects perceptual sensitivity. These heartbeat-related modulations are linked to fluctuations in interoceptive attention and to unconscious predictive coding processes within the brain. This research suggests that ongoing physiological rhythms—specifically cardiac signals—directly modulate perceptual sensitivity, contributing to an adaptive, yet largely unconscious, fine-tuning of sensory experience. For instance, they found that

> … the modulation of perception and neural response along the cardiac cycle, seems best explained by periodical modulations of perception in a predictive coding framework, in which the brain is continuously producing and updating a model of sensory input. This model not only concerns exteroceptive stimuli but also interoceptive signals such as the heartbeat. Each heartbeat and its concomitant pulse wave lead to transient physiological changes in the entire body. These repeating cardiac fluctuations are treated as predictable events and attenuated by the brain to minimize the likelihood of mistaking these self-generated signals as external stimuli.[41]

As Peter Dockrill explains, "we seem to perceive things less at the moment that the heart clenches and pumps blood through the body. Directly after, though, when the heart relaxes before the next heartbeat, we can register sensations more clearly." This highlights how bodily rhythms, including hormonal fluctuations and visceral or musculoskeletal processes, can disrupt and distort not only visual perception but also memory, focus, and decision-making.[42]

The underlying implication is that pre-conscious, bodily-driven elements of "information generation" can influence specific actions without

[41] Al et al. (2020: 1, 6).

[42] Damasio (1994); Gallagher (2005); Shapiro (2004).

necessarily aligning with the immediate task. Our awareness of these bodily and mental processes is fragmented, manifesting only as a visual or perceptual object. The "filling-in" function—often framed as an evolutionary mechanism prioritizing rapid response over deliberate analysis—enables immediate action without space for deliberation. This unconscious processing challenges theories that align propositional truth with visual perception, as it implies that much of perception operates outside propositional reasoning.

Meanwhile, dual-processing theories often conceptualize the mind as comprising multiple distinct systems. Though frequently interpreted through a cognitivist lens, these theories primarily emphasize brain functionality. Functionalist perspectives can, to a limited extent, serve as helpful abstractions, visualizing information processing as distributed yet coordinated without a singular controlling mechanism. However, an overreliance on functionality risks oversimplification, neglecting the complexity of cognitive processes.

A common simplification in cognitive science involves abstracting the "mind" as a non-physical and purely mental entity, positioned above and distinct from physical-physiological processes. Contemporary cognitive science, however, increasingly recognizes the brain (and mind) as an interwoven system of multiple, concurrent processes—neural, chemical, and physiological—that interact dynamically. This realization has led researchers to pivot toward frameworks that emphasize dynamic interactions and adaptive shifts among reasoning processes, like the theory of nonlinear dynamical systems, which integrates phenomenology and hermeneutics, acknowledging the complexity of mental processes as interdependent, fluid, and context-sensitive.

CHAPTER 3

# Cognitivism

Cognitivism, a dominant paradigm in cognitive psychology, seeks to understand mental processes through computational and representational models of the mind. While closely related, cognitivism and cognitive psychology serve distinct roles: cognitivism provides a theoretical framework for cognition, whereas cognitive psychology is an empirical discipline dedicated to studying mental processes. Despite their differences, both share a fundamental commitment to investigating how humans acquire, process, and apply information.

The intersections of phenomenology, hermeneutics, and cognitive psychology stem from their shared focus on human experience, meaning-making, and the underlying structures of perception and cognition. However, each field approaches these concerns through distinct methodologies and theoretical orientations.

Phenomenology, particularly as developed by Edmund Husserl, emphasizes first-person experience and the structures of consciousness. Husserl introduced epoché, or "bracketing," to suspend preconceptions and analyze phenomena as they appear to consciousness. The primary aim of phenomenology is to provide rigorous, descriptive accounts of lived experience, uncovering the fundamental structures of subjective awareness.

Hermeneutics, as advanced by thinkers such as Hans-Georg Gadamer and Paul Ricoeur, shifts the focus to interpretation and meaning, particularly within linguistic and cultural contexts. A core hermeneutic principle



J. K. B. O. Friis, *Hermeneutics at the Intersection of Medical Technology*, https://doi.org/10.1007/978-3-031-88755-0_3

is that understanding is always mediated by historical context and preconceptions. Unlike phenomenology, which seeks to describe experience in its immediacy, hermeneutics contends that all experience is interpretative, shaped by prior understandings. Thus, hermeneutics complements phenomenology by embedding experience within a broader historical and linguistic framework, emphasizing the role of interpretation in meaning-making.

Cognitive psychology, in contrast, employs empirical and experimental methodologies to investigate mental functions such as perception, memory, and problem-solving. Early cognitive psychology was influenced by information-processing models, which conceptualized the mind as a computational system. However, later developments, including embodied cognition and enactivism, have incorporated insights from phenomenology, recognizing the significance of bodily experience and situated action in cognition.

A crucial connection between phenomenology and hermeneutics lies in their shared critique of naturalism and their emphasis on lived experience and meaning. While phenomenology strives to describe experience as it presents itself, hermeneutics extends this analysis by asserting that all experience is shaped by historical and linguistic contexts.

The relationship between phenomenology and cognitive psychology has evolved through approaches such as embodied cognition, enactivism, and 4E cognition, collectively known as neurophenomenology. These perspectives integrate phenomenological insights to account for the role of bodily experience in cognitive processes.

A central critique from hermeneutics challenges cognitive psychology's reliance on computational models, arguing that cognition is not merely data processing but an interpretative process influenced by meaning, narratives, and context. This critique has spurred interdisciplinary efforts to refine cognitive science through hermeneutic and phenomenological insights.

Despite their methodological differences, phenomenology, hermeneutics, and cognitive psychology remain deeply interconnected through their mutual concern with experience and meaning. Phenomenology provides a foundational descriptive approach, hermeneutics offers an interpretative framework, and cognitive psychology seeks empirical validation of mental processes. Contemporary approaches, such as enactivism (Shaun Gallagher), neurophenomenology (Daniel Hutto), neuro-hermeneutics (Karl Friston), and embodied cognition (Mark Johnson), endeavor to bridge these disciplinary divides, integrating insights from all three traditions to advance a more holistic understanding of cognition.

The term "cognitivism" first emerged in 1956, following influential meetings among prominent figures such as Herbert Simon, Noam Chomsky, Marvin Minsky, and John McCarthy.[1] According to Livingston, cognitivism posits that human cognition operates through the computation of symbolic representations, describing this theory as a "formalist" approach to understanding the mind. As Livingston notes, the strength of cognitivism lies less in empirical validation and more in its "sophisticated appreciation of the logic and conceptual grammar of psychological terms."[2] This formalism contrasts with experiential or behavior-based theories, emphasizing the structural logic underlying mental functions.

One prominent, modern branch of cognitivism is teleosemantics, a theory of mental representation that defines mental states as reliably caused by specific stimuli and shaped by evolutionary or learned responses to these stimuli. Jesse Prinz describes teleosemantics as follows: "A mental representation is a mental state that is reliably caused by something and has been set in place by learning or evolution to detect that thing … a mental representation is a mental state that has been set up to be set off by something."[3]

Central to cognitivism is the assumption that human cognition is inherently structured to represent the world objectively and truthfully. In applied research, particularly in cognitive psychology studies related to radiology, scholars employ a variation of cognitivism known as psychophysics. Psychophysics emphasizes precise measurements and quantitative analysis, essential in tasks like calculating observer variability or using brain scans to track when and where perceptual phenomena, such as illusions, arise during image interpretation. This approach suggests that quantitative data can provide a stable framework for understanding cognitive functions, such as visual processing in the cortex. In this context, researchers measure electrical neuron responses to visual stimuli to classify cortical neurons that respond to specific image features, such as color or contrast. This selective process in the brain transmits visual information from the

[1] Varela et al. (1993: 40).
[2] Livingston (2005: 30).
[3] Prinz quoted in Hutto (2012: 2).

retina via cortical neurons to other brain areas for subsequent processing by memory, conceptualization, and motor planning.[4]

Cognitivism, however, contrasts sharply with embodied cognition theories, which assert that cognition is not confined to the brain but is distributed across the body—and sometimes extended into the external through technologies. Embodied cognition incorporates emotions as fundamental components, with the body's entire experiential and emotional systems influencing cognitive processes. Unlike cognitivism, embodied cognition rejects representationalism, suggesting that emotional and cognitive responses are not purely symbolic representations of the world but are holistic responses involving the entire organism.[5] Enactivist perspectives on emotion, for instance, emphasize that emotions are not solely cognitive attitudes or appraisals but are complex organismic responses. Daniel Hutto argues that to describe emotions as "cognitive" is misleading, as it implies that emotions are intentional mental states, such as beliefs or judgments, which can be revised through reasoning.[6] Hutto critiques this view by asserting that it overlooks the importance of bodily feelings in emotion; emotions do not function as propositional attitudes that represent situations in truth-evaluable ways.

Hutto further contends that emotions are better understood as responses to relational themes, involving distinctive bodily feelings rather than as cognitive states that "represent" reality. Emotions like anger, for example, do not inherently "represent" an actual offense. Instead, they reflect a "setup" of responses to relational themes, which trigger feelings without asserting claims about reality. Thus, the enactive perspective proposes that emotions engage the whole organism in ways that transcend mere symbolic representation, emphasizing the affective and embodied aspects of human experience.

Embodied cognition theories incorporate both chemical and electrical processes within the brain, distinguishing them significantly from traditional cognitivist frameworks, which focus exclusively on neural electrical activity. Embodied cognition theorists acknowledge the computational cognitivist emphasis on neural "electrical firing," yet they extend this view by emphasizing the role of chemical dynamics in brain function. This distinction echoes Alan Turing's insight from 1950, where he noted that

[4] Varela et al. (1993: 44).
[5] Hutto (2012: 6); Thompson (2007: 363).
[6] Hutto (2012: 1).

chemical processes are as vital as electrical ones in the nervous system.[7] Wheeler, navigating a nuanced position between cognitivism and embodied cognition, highlights that cognitivism's view is rooted in an abstraction of the brain's physical organization, treating it as a distributed network of processing units focused solely on electrical signaling. This abstraction excludes the chemical interactions between brain and body, which embodied cognition considers fundamental to understanding human cognition holistically.[8]

A common metaphor, "understanding is seeing," encapsulates the embodied approach by suggesting that cognition is deeply rooted in perceptual and bodily interactions. This metaphor, used to map experiential understanding (source) to conceptual understanding (target), reflects that human cognition involves perceptual, cognitive, and language-based processes embedded in bodily activity.[9] Mark Johnson, who worked extensively with George Lakoff, introduced the term "embodied cognition" and later described it as "second-generation cognitive science" in 1999.[10] Johnson critiques mainstream cognitive science for its persistent view of cognition as an isolated, brain-based function that disregards the integrated role of the body. According to Johnson, this Cartesian dualism fails to account for the dynamic, interdependent processes through which the brain and body co-construct experience, thought, and action, mistakenly treating the mind as an isolated faculty of abstract operations independent of physical embodiment.

Cognitivism, therefore, is often seen as a reductionist approach, preoccupied with computational and representational models of neural transmissions while neglecting the broader experiential dimensions of human cognition. This reductionism has shaped cognitive science's understanding of human experience in ways that overlook the integrative role of bodily processes. In their seminal critique, Varela, Thompson, and Rosch[11] argue that cognitivism's limitations lie in two primary areas. First, cognitivism assumes that cognitive processes are largely inaccessible to conscious awareness, resulting in a fragmented view of the human self and a gap between scientific accounts of cognition and lived experience.[12] This

[7] Turing (1950: 46); Wheeler (2014: 378).
[8] Ibid.
[9] Johnson (2017: 14).
[10] Ibid.: 17.
[11] Varela et al. (1993).
[12] Ibid.: 48.

framework suggests that cognitive functions are fundamentally removed from the realm of conscious experience.[13]

Varela and colleagues propose that, while certain cognitive processes may indeed operate at subconscious or pre-conscious levels, these processes are nonetheless integrated with bodily functions and affective states, influencing our situated, meaningful engagement with the world. This perspective diverges from cognitivism by viewing neural processing as inherently linked to sensory and bodily functions. In this embodied view, pre-conscious responses and enactions contribute to a layered process that eventually culminates in conscious awareness, where meaning-making emerges as a gestalt rather than a computational or representational operation alone.

Thus, embodied cognition posits that pre-conscious reactions—such as those generated by sensory input and bodily states—are not separate from conscious experience but form a continuum with it. This dynamic interaction between neural, chemical, and bodily processes provides a foundation for conscious awareness, illustrating the interdependent processes through which cognition is enacted rather than merely represented. In this way, embodied cognition offers a more holistic framework for understanding human experience, integrating pre-conscious, affective, and sensory processes as central to the emergence of conscious thought and meaning-making interpretation.

We experience conscious awareness of what occurs both within us and in our surroundings, yet this awareness is preceded by a layer of pre-conscious sensing—a level where we are affected in ways that are not yet fully enacted. By the time a perception reaches conscious awareness, it has already been imbued with layers of meaning that extend beyond the raw sensory input. It has acquired perspective, background, context, materiality, qualities, relevance, and potential for action. This pre-enacted meaning shapes our conscious experience, feeding back into the ongoing, dynamic exchange of neural, chemical, and physiological processes within the integrated mind-brain-body system. In essence, consciousness is part of a continuous, evolving process of sense-making, constantly reconfiguring our lived understanding within specific contexts.

This framework naturally reshapes how we view experience as memory. Rapid processing, instinctual responses, emotions, and long-term memory intersect and interact with short-term memory. Together, they allow us to

[13] Friis (2011a, 2011b, 2015a, 2015b, 2016b, 2020a, 2021).

adapt to the immediate contexts in which we find ourselves. Consciously, we are aware only of the meanings and, at times, the actions derived from these dynamic exchanges. However, we do not directly perceive the underlying neural processing, synaptic activity, or the way information from bodily chemical interactions is distributed to the brain and translated into a coherent sense of an event. Our knowledge of these processes relies on mediations, such as medical imaging, which allows us to observe simultaneous neural activity across diverse brain regions. These images reveal how sensuous stimuli can activate specific brain areas; however, this intricate processing lies beyond our conscious control, desire, or intention. Our perceiving bodies interact autonomously with the environment, essentially operating "on our behalf."

Freud's conception of mind and consciousness also recognized a division, introducing the concept of the "pre-conscious." Pre-conscious activity involves neural processing below the threshold of awareness, where sensory input is transformed into unconscious "sensory givens." This introduces a discontinuity within the cognizing individual: identity and self are not singularly coordinated entities but are embedded within interpersonal, group, and cultural structures. Identity is a dynamically coordinated, sociocultural construct without a fixed Cartesian ego—no central conductor orchestrates thought and action. Instead, sense-making unfolds in parallel, with certain processes occurring rapidly, drawing on long-term and short-term memory to maintain our awareness and enable adaptive responses to events as they unfold.[14]

Where my perspective diverges from cognitivism is in the assumption that cognition and representation are inherently linked.[15] Consciousness is not the origin or the "center" of cognitive activity, nor is it synonymous with cognition itself. Consciousness emerges as the result of embodied cognition; it reveals the outcomes of neural and chemical interactions that have already been processed. Put simply, we are consciously aware of what we see or perceive only after these processes have already brought it to awareness.

This stands in stark contrast to traditional cognitivism—not in the view that the brain actively constructs a representation of the world to which we have only limited perceptual access, but rather in the assumption that the brain's representations are not necessarily truthful or accurate

[14] Friis (2011a, 2011b, 2015b, 2020a, 2021).
[15] Varela et al. (1993:49); Friis (2020a, 2021).

depictions of external reality. Our thoughts, expressions, shared understandings, and self-conceptions do not inherently convey objective truths. Moreover, our meaningful experiences do not occur primarily in the service of truth itself.

The divergence between views here centers on two distinct processing systems: one that encompasses perception and automated enactments, and another that conceives cognition as conceptual understanding. For instance, in radiology, the interaction with an image involves a form of simulation that activates selective sensory and motor aspects of the image, alongside the typical physical and cultural interactions associated with it.[16] Thus, a meaningful understanding of human image analysis in radiology requires examining human-technology interactions, especially the interpretative character of embodied cognition within these exchanges. Our interactions with the environment, other people, and technology are interpretative processes, marked by an ongoing drive to ascribe meaning to the continuous flow of perceptions and bodily exchanges within our complex surroundings. This introduces a hermeneutic perspective to the "body-brain-mind-environment/technology" interaction.[17]

There appears to be a critical flaw in the cognitivist metaphor that likens the brain to a computer; in many ways, this view echoes the outdated mechanistic conception of humans as purely physical machines. Cognitivism's approach to the mind as a computational system effectively reduces the body to a mere "biomass machine" composed of muscles, joints, and tissue, with no bearing on cognitive processes themselves. This perspective aligns with that of MIT professor Rodney Brooks, who asserts, "The body, this mass of biomolecules, is a machine that acts according to a set of specifiable rules... We are machines, as are our spouses, our children, and our dogs... I believe myself and my children all to be mere machines."[18]

Brooks's statement underscores a purely mechanistic perspective, suggesting that all human actions and experiences can be reduced to predictable, rule-based processes. In this view, cognitive functions operate independently of the body's material and interactive context, effectively sidelining any influence of bodily experience or environmental interaction on cognitive processes. By maintaining that humans are "mere machines,"

[16] Johnson (2017: 23).
[17] Friis (2011a, 2011b, 2015a, 2015b, 2016a, 2016b, 2020a, 2021).
[18] Brooks, quote in Gallagher (2008: 141).

cognitivism neglects the dynamic and reciprocal relationship between mind and body—a relationship that theories of embodied cognition argue is essential for understanding the complexities of human thought, perception, and action. This mechanistic stance, then, may limit cognitivism's capacity to fully account for cognition as it unfolds within the body's lived, sensorimotor engagement with the world.

CHAPTER 4

# Historical Attempts at Embodied Hermeneutical Cognition

Interpreting is to fully immerse oneself in the present moment as an active participant. It involves noticing, observing, and deeply sensing what is unfolding. This mode of thinking is inherently embodied—rooted in our physical experience and inseparable from our being. As an interpretive process, it does not merely receive reality but engages with it. In this way, interpretation is a form of attunement, a state of readiness in which meaning emerges and reveals itself to consciousness.

In the context of radiological practice, this is how interpretive analysis operates: it is a situational, context-dependent process of interpretation and embodied perception. Practitioners must remain open, awaiting the emergence of patterns in the images they study. Here, thinking is not detached analysis but an interpretive, perceptual act shaped by the body's engagement with the concrete physical aspects of the situation. However, perception itself is always contextualized, infused with memories and past experiences that reframe the present.

This book draws on the ideas of thinkers such as Hume, Heidegger, Nietzsche, Gadamer, Varela, Merleau-Ponty, Zen Buddhist philosophy of mind, and Wittgenstein, as it seeks to explore the limitations of our cognitive capacity to grasp truth beyond ourselves, beyond society, culture, or any abstract absolute. My argument is that truth is not an external entity waiting to be discovered, but rather an invention generated within the frameworks of belief and normative systems. For too long, truth has been



J. K. B. O. Friis, *Hermeneutics at the Intersection of Medical Technology*, https://doi.org/10.1007/978-3-031-88755-0_4

treated as an arbitrary endpoint in philosophical and scientific inquiry, a stance that reflects deeper tensions within human nature itself.

The claim that truth exists "out there" independent of our interpretive frameworks is a position I want to challenge; instead, we invent truth rather than uncover it. Our realities, practices, and mental landscapes are constantly in flux, and it is within these evolving mindsets that truth takes shape. Thinkers like Hume, Nietzsche, and Wittgenstein acknowledge that we are not fully in control of our own cognitive faculties—we are not the "masters of our own house." The complexity and opacity of the world cannot be easily penetrated by human reasoning alone.

This epistemological dilemma is further highlighted by Kurt Gödel's Incompleteness Theorem, which demonstrates that the scope of truth is so vast and all-encompassing that there will always be questions that remain unanswered and perhaps unanswerable. No matter how much we think we know, our answers will always be incomplete,[1] underscoring the inherent limitations of human understanding. In this view, truth is not a fixed endpoint but an ongoing process shaped by our engagement with the world and the limitations of our cognitive and cultural frameworks.

The incompleteness of arithmetic logic became evident with the realization that it must rely on unprovable statements, a concept later formalized in Gödel's Incompleteness Theorem. However, the groundwork for this idea was anticipated as early as 1914 by a young Norbert Wiener, who published an article in the *Journal of Philosophy, Psychology, and Scientific Method*. Wiener remarked, "It appears to me unlikely… that we can get… certainty that is absolute in the propositions of logic and mathematics, at any rate in those that derive their validity from the postulates of logic."[2] This early insight foreshadowed Gödel's more formalized conclusions about the inherent limitations within logical systems.

Following this trajectory, Gödel's theorem can be linked to Wittgenstein's *Tractatus Logico-Philosophicus*,[3] first published in 1921 as *Logisch-Philosophische Abhandlung*. This work significantly influenced the Berlin and Vienna Circles, leading many of them to dismiss metaphysics and religion from rational discourse, based on Wittgenstein's last proposition in his *Tractatus*: "Whereof one cannot speak, thereof one must be silent." However, this interpretation was a profound misreading

[1] Gödel (1931).

[2] Quoted in Conway and Siegelman (2005: 33).

[3] Wittgenstein (1989).

of Wittgenstein's intention. Wittgenstein later criticized the logical positivists and empiricists for their misunderstanding, emphasizing that those things we cannot speak of are the most important things—the only ones that are truly important. Thus, Wittgenstein arguably suggested that logic and science alone are insufficient for a full understanding of the world; there are aspects of reality that transcend what can be articulated in factual or scientific terms.[4]

Furthermore, Wittgenstein delves into the nature of free will and the limitations of knowledge, asserting, "The freedom of the will consists in the impossibility of knowing actions that still lie in the future." He suggests that if causality were an inner necessity, we could predict future actions, but this is not the case. Additionally, he challenges the assumption that self-evidence guarantees truth: "The truth of a proposition does not follow from the fact that it is self-evident to us... its self-evidence in no way justifies our belief in its truth."[5] Wittgenstein's reflections highlight the deep uncertainty that permeates both logical propositions and human understanding, underscoring the limits of what can be known with certainty.

Another significant "leap"—in my opinion—occurs in the work of Austrian writer and philosopher Robert Musil, particularly in his novel *The Man Without Qualities*. Musil's protagonist reflects on the nature of modern sciences, offering insights that align closely with Wittgenstein's philosophical views. He observes how analysis dissects even the simplest human actions into layers that span across millennia, remarking, "the needs of life are different from the requirements of thought," and that what happens in life often contradicts the patterns to which a trained, rational mind is accustomed.[6] This tension mirrors the conflict between two modes of human existence: the subconscious, bodily processes that govern much of our daily functioning, and the rational, calculative thinking instilled by education.

On the one hand, we rely on subconscious acts—complex motor-sensory, neurochemical, and physiological processes—to navigate daily tasks, most of which we perform without conscious deliberation. On the other hand, formal education installs within us a more detached, logical mode of thought designed to curb the irrational forces that millennia of

[4] Doxiadis and Papadimitriou (2009: 258, 259).

[5] Wittgenstein (1989: 39).

[6] Musil (1995: 329).

human evolution could not fully eliminate. Nietzsche addresses this friction when he writes, “Behind your thoughts and feelings, my brother, stands a mighty commander, an unknown wise man—his name is Self. In your body he dwells, he is your body.”[7] This “wise” Self acts based on what it already knows, guiding us instinctively, like a shadow following our every move.

Humans, shaped by evolution, are equipped to respond to incomplete sensory data, circumventing the time lag between the perception of an event and our conscious awareness of it. This ability allows us to act seamlessly in real time, even in complex environments. Whether riding a bicycle, managing pandemics, or exploring space, we have adapted to situations that far exceed the capabilities of our conscious, logical reasoning. Science, in contrast, demands precision, objectivity, and accuracy—qualities that expose the inherent limitations of human nature. Modern neuroscience exemplifies this tension, showing how our minds and bodies often operate beyond the boundaries of what rational, conscious thought can fully grasp. Despite our striving for scientific exactitude, we remain intimately bound to our imperfect, evolved nature.

Grigori Guitchounts,[8] a young Russian-born neuroscientist with a doctorate from Harvard University, observes that while significant strides have been made in understanding brain architecture and certain aspects of brain function, much remains unknown. Neuroscientists have identified regions of the brain that respond to environmental stimuli, activate our senses, and generate movements and emotions. Yet, as Guitchounts notes, we still lack a comprehensive understanding of how these different regions interact and depend on one another, and how these interactions give rise to complex phenomena like behavior, perception, and memory. The processes through which the brain creates memories, thoughts, perceptions, emotions, and consciousness itself remain elusive, hidden within the intricate web of neural connections.

Today’s neuroscientists utilize advanced technologies like 61-beam electron microscopes to visualize the minutest details of brain tissue, combined with machine learning to map the brain’s architecture. Despite these technological advances, Guitchounts raises an essential question: where does scientific inquiry end and personal interpretation begin? His answer suggests that many scientific judgments are ultimately intuitive, and thus

[7] Nietzsche (2011: 56).

[8] Guitchounts (2020).

subject to variation. One prevailing intuition in neuroscience is that neurons are extraordinarily complex entities. They exhibit back-propagating action potentials, possess dendritic compartments that operate independently, and are equipped with multiple channels, leading some to propose that a single neuron might function as a network unto itself.

Like many scientists and philosophers before him, Guitchounts once regarded science as the sole path to knowledge. However, he has since relinquished the ideal of scientific objectivity, recognizing that science is fundamentally an interpretive endeavor. According to Guitchounts, facts are always filtered through an interpretive lens, even before experiments begin. The very act of choosing what experiments to conduct, and how to design them, is shaped by human biases and subjective decisions. This interpretive process continues after data collection, when scientists must decide how to make sense of the data. Thus, while science gathers facts about the world, it is humans who interpret and describe these facts. As Guitchounts argues, scientific understanding is always filtered through a personal and cultural sieve, meaning that the pursuit of knowledge is as much an act of interpretation as it is of discovery.

We have created science and technology, yet we have come to recognize that we may not be the most reliable observers for collecting or analyzing data, especially when it comes to speed, precision, and granularity. As discussed earlier and argued later, our biological nature is inherently adaptable, shaped by bodily and cognitive processes that often lie beyond our control.

The trained perspective—such as that of a radiologist—is, in many ways, an imposed one. The challenge with this trained perspective arises from how our minds have been shaped both by evolution and by the cultural contexts we inhabit. Through evolution, the brain-body-mind system has been wired for survival, acting instinctively and rapidly in response to environmental cues. This evolutionary wiring allows us to bypass the temporal constraints of conscious, causal processing, enabling us to act on fragmented sensory information.

Human nature is thus a product of both natural and cultural forces, with each dimension pulling our perceptual abilities and cognitive skills in different, often conflicting, directions. While nature equips us with instinctual responses, culture overlays these with learned patterns of thinking and interpretation. These interpretations, however, are not precise reflections of reality but rather inductive approximations. They are shaped by the limitations of our biology and the demands of the cultural environments we navigate. Consequently, our understanding of the world is not a straightforward product of objective observation but is filtered through layers of evolutionary and cultural conditioning.

CHAPTER 5

# Interpretation

Interpretation is not confined to a single method of object-analysis; rather, it encompasses a diverse range of approaches across different domains of experience. This plurality of interpretive methods stems from a fundamental shift in philosophical hermeneutics, which paralleled the transformation of scientific thought in the nineteenth century. During this period, long-established idealist philosophical systems came under scrutiny, prompting a broader intellectual upheaval that reshaped not only philosophy but also the sciences. In response, a movement emerged advocating for the autonomy of individual scientific disciplines while fostering a dynamic interplay between philosophy and these fields. Heidegger was a vocal critic of the era's dominant positivist and empiricist philosophies, which sought to universalize the research methods of the natural sciences. He condemned this prevailing perspective as a reductive materialism—one that stripped inquiry of its depth by subordinating all understanding to the framework of the natural sciences.

Medical science and its practices once epitomized this materialistic or positivistic approach, but the landscape has since changed. The integration of complex technologies now demands a deeper engagement with the meaning embedded in its mediations. Clinical practices, such as cardiology and radiological image analysis, exemplify modern interpretive processes that transcend mere empirical observation, reflecting a broader interpretative practice.



J. K. B. O. Friis, *Hermeneutics at the Intersection of Medical Technology*, https://doi.org/10.1007/978-3-031-88755-0_5

## 5.1 The Making of an Interpretative Practice

Hermeneutics itself, originally conceived in the sixteenth century, began as a specialized approach to interpreting texts, particularly religious scriptures and classical writings. The goal of early hermeneutics was to establish a methodical framework for attaining true understanding of these texts. Central to this approach was the concept of the hermeneutical circle, a process by which interpretation moves between the individual parts of a text and its overall meaning. In classical hermeneutics, each part of a text is understood in light of the whole, and the whole in turn is clarified by the parts. Through iterative engagement with this circular process, interpreters aim to arrive at a comprehensive understanding free of internal contradictions. The validity of an interpretation is strengthened when all parts of the text cohere with its general meaning, ensuring a consistent and holistic understanding.

The transformation of philosophy and science in both the nineteenth and twentieth centuries can be traced in part to the influential works of Friedrich Schleiermacher (1768–1834), Wilhelm Dilthey (1833–1911), and later, Emilio Betti (1890–1968). These philosophers played a pivotal role in establishing hermeneutics as a legitimate scientific method. Schleiermacher and Dilthey, in particular, viewed individual texts not as isolated works, but as expressions of an author's broader body of thought, which is inseparable from the author's life, societal context, and historical epoch. This approach situates both authorship and interpretation within the larger framework of world history. According to Schleiermacher, Dilthey, and Betti, hermeneutics should be a rigorous, analytical methodology applicable across the humanities and social sciences. Dilthey, in particular, aimed to elevate hermeneutics to the same level of validity as the methods of the natural sciences by providing a logical, epistemological, and methodological foundation for the human sciences.[1]

With Schleiermacher and Dilthey, the "hermeneutical problem" evolved into a scientific challenge. This problem arises from the recognition that interpretation is shaped by the complex interplay of factors such as language, tradition, worldview, and the intellectual currents of the time, all of which influence the interpreter.[2] Unlike the objectivity sought in the natural sciences, hermeneutics acknowledges that interpretation is

[1] Grondin (2005: 984).
[2] Ricoeur (1974: 5).

contextually embedded, making it a more intricate process. It requires not only cognitive understanding but also psychological faculties such as empathy, allowing the interpreter to immerse themselves in the life and psyche of others. The method seeks to establish interpretation as a logical and epistemologically sound process, countering the notion that it is a relativistic or purely subjective endeavor. Schleiermacher and Dilthey thus introduced a more comprehensive view of interpretation, one that integrates personal experience, scholarly knowledge, and psychological insight, distinguishing it from the quantitative methods of the natural sciences.

Schleiermacher and Dilthey elevated hermeneutics beyond mere interpretation, assigning it a normative and regulatory role in understanding written texts across various disciplines. They developed specialized hermeneutical methods tailored to distinct academic fields—for example, *hermeneutica sacra* for theology and sacred texts, *hermeneutica juris* for legal interpretation, and *hermeneutica profana* for classical literature. Central to these methodologies was the hermeneutical circle, the dynamic interplay between parts and the whole. However, true comprehension of a text requires more than this structural approach. Interpreters must also account for its genre, examine the author's broader body of work, and situate the text within the author's personal history and the historical context in which it was created.[3]

Heidegger acknowledges that Dilthey's hermeneutics was often misinterpreted superficially. Heidegger himself was profoundly influenced by Dilthey's insights, particularly the emphasis on cultural and historical influences in shaping human understanding. Heidegger critiques the tendency of scholars to rigidly separate the natural sciences from the humanities, imposing distinct rules on disciplines like history and psychology, which led to a relativistic "philosophy of life." While this interpretation captures some aspects of Dilthey's thought, Heidegger argues that it misses the deeper essence of the matter.[4]

Heidegger further expands on Dilthey's concept of the hermeneutical circle, describing it as a process of "trying things out," driven by a fundamental restlessness, which he sees as inherent to human nature. In Heidegger's view, the ultimate aim of humanity is to understand life itself. Dilthey's research sought to distinguish the methodologies of the humanities from those of the exact sciences, focusing on the historical

[3] Grondin (2005: 984).

[4] Heidegger (1992a), *Being and Time*, 450.

development of knowledge and the interpretation of scientific thought. To fully grasp the mindset of the interpreter, however, Dilthey realized that it was essential to explore human psychology. His work revealed that the interpretation of history, scientific inquiry, and psychological understanding are deeply intertwined. As Heidegger states, "psychology is the central focus" because it enables an understanding of life in its historical context, providing insight into the essence of human existence. Through this lens, Dilthey's hermeneutics goes beyond mere interpretation, becoming a means of comprehending human life in its fullness.[5]

## 5.2 Understanding as Process

Understanding as an eventful process highlights the notion that gaining knowledge and comprehension is not a static or passive experience but an active, evolving interaction that involves engagement and reflection. Philosophers such as Martin Heidegger and Hans-George Gadamer underscore the intrinsic link between life and thought, positioning understanding as a fundamental dimension of human existence. From a hermeneutic standpoint, understanding is both a methodology for interpreting texts, images, and other forms of knowledge, and a deeper process of existential attunement. As Heidegger suggests, interpretation is a dynamic unfolding that goes beyond the mere acquisition of information. Instead, it attunes individuals to the world, enabling the enactment of meaning through lived experiences.

Hermeneutics, therefore, transcends being a mere methodology. It encompasses the broader scope of human cognition and perception. Often recognized as a method for interpreting cultural artifacts—such as elegies, paintings, architectural structures, or symphonies—hermeneutics involves elaborating on the perceived content, possibly in a cultural context that extends beyond the author's or composer's original intent. At its core, however, hermeneutics is not just about interpreting human-made things; it is a theory of interpretation rooted in the lived world. It represents the very essence of human cognitive activity, permeating all aspects of what we do, think, and perceive.

[5] Ibid.: 451.

In the Continental philosophical tradition, particularly within German and Scandinavian contexts, epistemology (*Erkenntnislehre*) is grounded in the interplay between cognition (*Erkenntnis*) and a formal methodological theory of knowledge. Cognition encompasses the processes of perception and thinking, operating across various mental states of pre-conscious and conscious awareness. Hermeneutics, understood as the pure act of interpretation, is inherently embodied, making it a fundamental activity within any cognitive-epistemological process. From its implicit beginnings in perception to the more abstracted realms of analysis, hermeneutics guides the development of understanding (*Erkenntnisvermögen*). In essence, it investigates how our inherent capacity for interpretation is embedded within our perceptual processes.

This interpretive capacity is shaped by embodied learning, practical expertise, lived experience, temporality, evolutionary instincts, cultural frameworks, traditions, and professional or familial influences. These factors collectively contribute to what Heidegger, Gadamer, and Ricoeur describe as the historicity of human understanding—what Gadamer refers to as our "prejudices" when examined through methodological scrutiny. Gadamer emphasizes the necessity of engaging systematically with our historicity, encapsulated in his well-known assertion: "It is always a past that allows us to say, 'I have understood.'"[6]

Perception itself is inherently interpretive, especially when we recognize familiar objects or concepts. Identifying something as a specific entity—such as diagnosing a blocked artery—demonstrates how understanding is always shaped by prior knowledge and contextual awareness.[7] Heidegger argues that human beings do not passively receive meaning but actively project it onto their experiences. We interpret the world through the lens of our own possibilities, understanding ourselves as beings oriented toward self-development.[8] This act of projecting meaning is, fundamentally, interpretation.

Crucially, we never encounter a world devoid of meaning; rather, we continuously impose structure and significance onto our experiences. Consider a medical procedure like angioplasty: while a cardiologist can diagnose a blocked artery, the interpretation of the best course of action requires integrating knowledge, skill, and situational awareness.

[6] Gadamer (1976: 58).
[7] Heidegger (1992a), *Being and Time*, 192.
[8] Ibid.: 188.

Understanding, then, is not merely about identifying facts but about actively engaging with and making sense of complex realities.

In the medical procedure of angioplasty, the task is framed by a clear objective—to address a blocked artery—but the pathway to its resolution is only partially revealed at the outset. The cardiologist's work is not a simple execution of predetermined steps but an active process of discovery and interpretation. This process involves guiding a catheter through the patient's artery, often from the arm to the heart, and requires a finely tuned reliance on tactile perception.

Unlike purely technical operations driven by visual feedback or fixed protocols, angioplasty is marked by a dynamic engagement with the specificities of the patient's unique anatomy. The cardiologist's tactile perception serves as a medium of "unveiling" the artery's condition, a process that is fundamentally hermeneutical. As the catheter moves through the artery, each moment of resistance, smooth passage, or unexpected friction communicates critical information. These tactile sensations are not neutral data points but meaningful signals that shape the cardiologist's ongoing interpretation of the artery's state.

The process can be understood as an act of appropriation, in which the cardiologist actively integrates sensory input into her evolving understanding of the situation. Initially, she operates with a general pre-understanding of the artery's possible conditions—for instance, its expected shape, elasticity, or degree of blockage. However, as she encounters new tactile feedback, this pre-understanding must shift. The artery is not a static object to be "read" but a dynamic context that reveals itself through action. Each movement of the catheter discloses new particularities, requiring constant reassessment of the strategy for navigating the vessel.

This interpretive engagement underscores the hermeneutical nature of medical practice. The cardiologist's hands do not merely relay technical information; they mediate a lived experience in which the hidden aspects of the artery are gradually brought into understanding. Each tactile cue is part of a larger interpretive process, where sensory data are not passively received but actively discovered within a shifting horizon of meaning. The artery's condition is therefore not something fully known in advance but something that is revealed step-by-step through the practice of skilled tactile perception.

Thus, angioplasty exemplifies a mode of problem-solving that resists reduction to algorithmic logic. It is not only the blockage that must be overcome but also the "opacity" of the body itself. The body does not fully disclose its nature from the start; it must be engaged, explored, and interpreted. In this sense, the cardiologist's work mirrors broader phenomenological insights into human experience. Just as meaning in a text is not fully given but unfolds through interpretation, the anatomical reality of the artery becomes knowable only through a process of ongoing interaction. The cardiologist's role is therefore not simply to "find" the solution but to actively co-create it through embodied practice, sensory interpretation, and conceptual refinement.

Heidegger's concept of the hermeneutical circle is particularly relevant here. Entering the circle "the right way"[9] involves recognizing the fore-structure—those preconceptions and prior understandings that guide interpretation—and allowing the process of interpretation to unfold in response to the specificities of the situation. Meaning is not imposed from outside but revealed through the interpreter's engagement with the world.[10] In this case, the cardiologist's understanding emerges from the interaction between prior knowledge and the evolving demands of the medical procedure.

Gadamer articulates a central concern of his philosophical inquiry by emphasizing that his focus is not on human actions or moral prescriptions, but on what happens to us beyond our intentions and deeds. The key challenge for the natural sciences, he suggests, lies in overcoming subjectivity to access an objective reality that exists independently of human perception. However, in the hermeneutical tradition, as represented by Heidegger and Gadamer, the goal is not to transcend subjectivity but to embrace it. As the later works of Heidegger, Grondin, and Gadamer demonstrate, the hermeneutic approach acknowledges that human existence is inextricably linked to subjectivity. Unlike the sciences, which strive for objective knowledge that exists apart from human cognition, hermeneutics engages with a world that is inseparable from human beliefs, perceptions, and experiences—a world in which we live, think, and ultimately die.

Heidegger's concept of *Dasein* encapsulates this idea, emphasizing that humans are always already immersed in the world. We do not stand apart from it; rather, we belong to and are shaped by it. Heidegger refers to this condition as being "thrown into the world," meaning that we find

[9] Heidegger (1992a) (Being…): 195.
[10] Ibid.

ourselves within a specific context that influences our understanding and shapes our experience. Through this immersion, our understanding is continually awakened and shaped by life and learning. *Dasein*, then, is humankind's engagement with its own Being, a fundamental concern with understanding our existence.

For both Heidegger and Gadamer, all understanding is historically situated. We are positioned within a tradition that reflects the thinking of people throughout time, and this tradition creates a constant fusion of past and present understandings. Gadamer refers to this process as the "fusion of horizons," a concept that, along with his idea of "historically effected consciousness," highlights the way in which our present understanding is shaped by the influence of historical ideas and cultural contexts. Tradition is not static; it is dynamic, constantly evolving as new interpretations and perspectives emerge through the interplay of past and present. Thus, to fully grasp the nature of human understanding, we must examine the role of tradition, culture, and history, recognizing that our comprehension is always informed by the legacy of those who came before us.

In *Hermeneutics and the Human Sciences*, Ricoeur compellingly describes "historically effective consciousness" as a pervasive and fundamental reality, wherein consciousness, even before it becomes fully self-aware, is already shaped by and dependent on the forces that influence it.[11] History, culture, and tradition are not merely external factors but are deeply embedded in our everyday lives, influencing our social, professional, and familial interactions. We encounter our historical and cultural legacies in our surroundings—through philosophy, literature, art, and conversations with others who may profoundly shape our thinking and understanding.

An illustrative example of this transformative interplay is found in the development of scientific thought. Over time, science has undergone significant paradigm shifts, which have altered both our ways of knowing (epistemology) and the nature of what we claim to know (ontology). These shifts within the natural sciences are often propelled by interpretations that reshape the object of inquiry, influenced by the interests of scientific traditions, cultural contexts, and even the personal perspectives of individual scientists. In the framework of modern hermeneutics, we do not claim access to an objective reality as it exists "in itself," independent

[11] *Italics* are mine.

of the perceiver. Rather, it views science, like all human activity, as activities inseparable from cultural and social life. While natural sciences may aspire to distance themselves from such influences, hermeneutics asserts that these cultural factors are essential in shaping interpretation and understanding—also within the natural and medical science practices.

Hermeneutics emphasizes that through language, we share and refine our interpretations of the world—whether they pertain to events, texts, images, or ideas. According to Gadamer, language serves as an intersubjective medium through which common understanding is generated. He uses the concept of "translation" to describe the process of interpretation, framing it as an act of translating meaning from one perspective to another. In this sense, understanding is an active engagement, in which the meanings embedded in the original language or context are transformed into our own way of expressing and comprehending them. This process reflects the "fusion of horizons," where the perspectives of both the interpreter and the author or speaker come together.

Crucially, this fusion does not mean that the interpreter simply reproduces the author's intended meaning. Rather, interpretation involves surpassing the original meaning, incorporating the interpreter's own horizon of understanding. Yet, this does not imply a complete divergence of ideas; on the contrary, language, tradition, and culture provide a framework that ensures a degree of alignment in the fusion of meanings. Through this shared cultural and linguistic medium, our individual interpretations are able to converge, enabling a mutual understanding that transcends mere subjective experience.

As will be argued later, this convergence of perspectives among medical professionals—such as cardiologists, radiologists, and forensic investigators—is essential to the diagnostic process. Each specialist brings a distinct yet complementary mode of analysis, shaped by their expertise, methodologies, and disciplinary frameworks. *Collaborative interpretation* emerges as a dynamic process in which these professionals integrate their insights, refining diagnoses through dialogue, shared observation, and the synthesis of multiple perspectives. This interplay is particularly critical in complex cases where no single viewpoint can fully capture the intricacies of a medical condition or forensic finding. Through this collective engagement, diagnostic accuracy is enhanced, uncertainty is reduced, and more effective treatment or investigative strategies can be developed.

CHAPTER 6

# Hermeneutics and Meaning-Making in a World of Things

In the last twenty years, there has been a noticeable increase in dissatisfaction among scholars in the field of continental philosophy. They are fed up with the endless interpretations and abstract ideas that dominate philosophical discussions. Instead, there is a growing push to focus on the actual, physical world and the importance of real, tangible things. Essentially, this means a significant shift toward realism within the realm of continental philosophy.[1] Günter Figal and Don Ihde belong to a group of thinkers within phenomenological-hermeneutical philosophy who have been reorienting hermeneutic philosophy toward the external origin of that *something* appearing in experience.

Both Ihde and Figal have significantly advanced hermeneutics, expanding its scope beyond the confines of hermeneutical "self-understanding" as emphasized by Heidegger and Gadamer. Figal, in particular, articulates a change by asserting that the hermeneutic object has now attained genuine external significance—however, Figal's realism recognizes how reality appears to the perceiver.[2] Figal strives to, in the words of Hans Ruin, "forge a hermeneutic philosophy that can be (…) faithful to the genuine thing-ness and exteriority of being, and that can also move beyond idealist

[1] Ruin (2019a, 2019b: 97).
[2] Ibid.: 100.



J. K. B. O. Friis, *Hermeneutics at the Intersection of Medical Technology*, https://doi.org/10.1007/978-3-031-88755-0_6

and subjectivist tendencies in phenomenological thinking."[3] This transformative aspect of hermeneutics, as proposed by Figal, is intricately linked to the objective nature of the subject under examination: interpretation is always an interpretation of *something*[4]—a concept that resonates with Ihde's material hermeneutics.

Despite the absence of a direct correlation between their respective writings, a nuanced exploration suggests a potential convergence between Figal's object-grounded hermeneutics and Ihde's material hermeneutics. In this light, both thinkers emphasize the interpretive act as a *mediation* of the subject matter, illustrating a shared focus on the substantive content of philosophical inquiry. The trick is, as Hans Ruin points out from a deep reading of Nietzsche, is not to "believe we have miraculously discovered how to reach outside and beyond thought and subjectivity simply by committing ourselves to a concept of the 'positive,' 'real,' or 'objective.'"[5] Nor, Ruin continues, to retreat to earlier forms of idealism—which explains why Nietzsche—trying to balance these temptations—could reject both idealism and realism.[6] Nietzsche ridiculed those who called themselves "realists" and who were claiming that the world is the way it appears to them—as if the real stood naked and pure in itself before them. This brings to mind the modern realist theories of representation—the mind being able to have direct access to reality and at the same time being able to keep in memory a truthful corresponding representation of it. Ruin quotes Nietzsche "You are still burdened with those estimates of things that have their origin in the passions and loves of former centuries. Your sobriety still contains a secret and inextinguishable drunkenness."[7] Even those believing they are incapable of being intoxicated by the belief that one have accessed the absolute have no access—there is no reality for these, nor for us, Nietzsche claims. It is simply impossible. Mediation is simply inevitable.[8] We access a mediated world where *some-thing* is interpreted.

[3] Ibid.: 102.
[4] Figal (2010: 56).
[5] Ruin (2019a, 2019b: 104).
[6] Ibid.
[7] Ibid., 105—Ruin's Nietzsche quote is from *Die fröhliche Wissenschaft*, KSA 3, p. 421.
[8] Ruin (2019a, 2019b: 106).

## 6.1 Mediation of Facticity

Hermeneutics is experiencing a profound shift toward a more holistic understanding of the world. This evolution moves beyond the confines of traditional text-centered analysis, embracing a wider framework that integrates insights from emerging developments in human-technology interaction. Within this framework, interpretation serves as a tool to uncover aspects of phenomena that may not be readily apparent. The mediation process plays a crucial role in maintaining the integrity and authenticity of the interpreted subject, while also situating it within the context of contemporary society to illuminate its relevance and significance. Figal recognizes interpretation as mediation that embodies understanding. Figal's understanding of interpretation as mediation goes back to the notions of the hermeneutics of facticity of both Heidegger and Gadamer.

During the year 1923 Gadamer is following Heidegger's lecture series on "Ontology," and, according to Figal, is reading Heidegger's *Natorp-Bericht*, where Heidegger for the first time speaks of a "hermeneutics of facticity." This becomes programmatic for Gadamer. Hermeneutics is brought together with "factual life."[9] Heidegger posits that understanding is an enactment of meaning.[10] For Heidegger, understanding is an inherent capacity with a projective nature, directing its focus toward the world and the potentialities in situations that we encounter.

Heidegger asserts, "Facticity… is the signification of 'our' 'own' Dasein."[11] In other words, our understanding of the world and ourselves is inseparably bound by the structures of "being-in-the-world." These conditioning structures shape the range of possibilities available in any given situation. For instance, a mood may determine our attunement to a situation, often in ways that are beyond our intentional control. Heidegger designates these conditioning structures as "facticity."[12]

Crucially, self-interpretation is not a process that occurs in isolation; it is necessarily intertwined with an ongoing cognition of the world. Our understanding of the world informs, and is in turn shaped by, our interpretation of ourselves.[13] Thus, the world is not a neutral backdrop for human existence but an essential component of self-understanding. Each

[9] Figal (2010: 8–9).
[10] George (2020: 1).
[11] Ibid.: 10.
[12] Heidegger (1992a). *Being and Time*, §§ 28–34.
[13] Ibid.: § 43; Fløistad (1987: 97).

individual's relationship with the world becomes a vital constituent of their self-interpretation, underscoring the inextricable link between self and world in Heidegger's existential analysis of Dasein.[14]

The concept of the human lifeworld encapsulates the interconnectedness of our self-perception and the external environment. This interconnectedness finds expression in the narratives we construct about our identity and actions. Essentially, we define ourselves through our experiences, thoughts, deeds, and emotions, some of which involve interactions with other individuals and entities. This encompasses a broad spectrum of activities, ranging from practical applications of objects to interpersonal communication and engagement with natural phenomena. This amalgamation forms our *Lebenswelt*.

Within our lifeworld, we coexist with both the natural world, with its objects and phenomena, and a cultural realm that encompasses everyday items such as X-ray machines, doorknobs, keycards, and mobile phones. Our lifeworld is a composite of personal experiences, the experiences of others, diverse cultural perspectives, the everyday utilization of objects, and the knowing-how of the scientific elucidation of natural phenomena. Consequently, it signifies the inseparable connection between humans and nature, emphasizing that human orientation extends beyond the self from the very inception of existence. In essence, we exist within a world, and our lifeworld is a manifestation of this coexistence.

The technologies that permeate our daily lives, including the various objects we employ, serve as tools for the meaningful care and organization of our existence within this world. These technologies not only facilitate our practical needs but also contribute to the structuring of our immediate environment. In this way, our lifeworld shapes and organizes by incorporating everyday technologies—a phenomenon underscoring our ongoing engagement with the world around us. For us humans, thrown into each our ways of existing, the factual is the presentness comprising our lives. Thus, the hermeneutics of facticity for Heidegger is not the art of interpretation but an articulation or enactment of life.[15] Hermeneutics is an exceptional "'how' of the 'ontological character of facticity'," Figal writes, it is a realization of facticity itself—"it is the illumination of a being that is in itself illuminated."[16]

[14] Ibid.
[15] Figal (2010: 10–11).
[16] Ibid.: 11.

Gadamer's concerns are rather different from Heidegger's notion of "facticity." In Gadamer, hermeneutical *truth* has center stage. This is not propositional truth but the hermeneutical experience of truth.[17] In the world of objects, this means a concern that extends beyond human creations to the world of things and our participation in the enactment of grasping and understanding these objects. Any experience of external objects is an interpretative event, and when the interpretation is successful, it is true. In other words, a radiologist, for instance, when looking at an X-ray image on the PACS identifies a malignant tumor by perceiving and identifying a pattern in the image which is similar to the patterns of tumors already experienced and memorized and as such it is an identification gestalted or established in consciousness as a true (plausible) diagnosis. The gestalt is here an enactment enabling memorized similar patterns with the visually perceived in an identification resulting in a true or plausible correct diagnosis (malignant cancer). Yet, the establishment of truth demands more than mere interpretation. As Gadamer posits, truth is attainable, albeit minimally, through the dynamics of hermeneutical conversation.[18] In this view, truth becomes a shared endeavor, a mutual concern among those engaged in the interpretative process. The diverse experiences, backgrounds, and expertise that each participant brings to the conversation are not obstacles but rather essential components in addressing the problem at hand. These differences enable the subject of interpretation to be viewed through various lenses, each shaped by distinct contexts.[19] Truth, then, is not an abstract or fixed entity but the outcome of discerning the most fitting solution—the optimal mediation—within the concrete situation before us. It emerges from the collective effort to resolve a problem through dialogue, where the plurality of perspectives contributes to a richer, more nuanced understanding.[20]

This process can be described as a "transformation into structure," where we grasp meaning and recognize it as true. Interpretation, in this sense, serves as a mediation. Meaning gradually comes into focus through

[17] Gadamer's concept of the *hermeneutical experience of truth* is unlike the traditional notion of truth as objective, verifiable, and independent of human experience, Gadamer's notion of truth is about something that emerges during the interpretive process—particularly in the context of human experience, history, and tradition.

[18] George (2022: 199–200).

[19] Ibid.

[20] I will give this notion of minimal truth a more concrete example later when we take another look at angioplasty.

repeated interpretative attempts, which may initially be inadequate, until the mediation becomes sufficient for understanding. For instance, a diagnosis represents a hermeneutical truth that remains open to further reinterpretation.

Rejecting naïve realism, metaphysical realism, indirect realism, and even postmodern or radical constructivist views, this perspective asserts that our interpretations actively shape what we call reality. However, this position also acknowledges that humans do perceive real objects in the world. Interpretation and understanding are inherent possibilities of human existence, rooted in the world itself. Yet, in the act of interpretation, we sometimes confront the opacity of objects, leading to misinterpretations. Human cognition and perception, while essential, have limitations that we must recognize and navigate.

The world does not consist of mind-independent entities, for human beings, along with animals, possess brains, minds, and sensory faculties. Thus, human interpretations and actions are part of reality itself. The space we inhabit positions us in referential relationships to things around us. Animals, too, perceive, act, and respond. The contents of the world are not mere mental representations, nor do they exist as inaccessible "things-in-themselves" that we can never fully grasp in their pure essence. Instead, reality emerges through the interplay of perception, interpretation, and action, shared by both humans and other sentient beings.

CHAPTER 7

# Hermeneutics and Objectivity

"The word itself says what an object is," Figal states.[1] It is what distinguishes between over and against, what is across from remains standing "across from." These oppositions allows for human expression: with precision we can determine something *insofar as* it is "over against."[2] Figal extends Gadamer's emphasis on the factual nature of interpretations within his hermeneutical realism framework. Figal's endeavor is externally focused; he prioritizes objectivity and real-world elements beyond the interpreting individual. We perceive of real-world objects. "An object is something standing in the way."[3] Consequently, Figal broadens Gadamer's contextual scope beyond art and human expressions. Interpretation, according to Figal, entails engaging with the objective reality of phenomena. Our understanding relies on external stimuli confronting us. Interpretation, therefore, acts as a conduit, facilitating the transmission of external elements that are foreign to us.[4] What places us in referential relation to the objects is, according to Figal, *space*. This referential relation to things is what generates interpretative possibilities—for us to make sense of them.[5]

[1] Figal (2010: 107).
[2] Ibid.
[3] Ibid.
[4] George (2009).
[5] Ibid.



J. K. B. O. Friis, *Hermeneutics at the Intersection of Medical Technology*, https://doi.org/10.1007/978-3-031-88755-0_7

We engage in interpretation as we experience the sensory impact of an external world filled with objects and are affected by the events that unfold within it. This access to a world beyond ourselves establishes a "mediate relation" to the world. Interpretation is a mediation of the thing, the occurrence—the matter at hand. As mediator the interpreter translates or negotiates—he brings the standpoint of one home to another, as Figal puts it.[6] Interpretation is also a recognition of something, a recognizing, which is a *carrying over* of the other. It is not as in Nietzsche's interpretation of *carry over*, where it becomes synonymous with metaphor, which concerns the leap from one sphere to another, from one metaphor to another metaphor—the world is a mixture of interpretations. Rather, *carry over* in Figal's hermeneutics has to do with something one make one's own (that is *carried over*) and yet is recognizable as something other, it can be carried over into a new context, and it should be traceable back to its extraction.[7] The concern of the interpreter is specifically with the matter.[8]

Familiar routines are often carried out automatically, without conscious thought. These actions are driven by neural and bodily interactions that work together to create meaning and perception—prompting the necessary response, whether it's opening a car door or identifying a tumor on an X-ray. According to Figal, we are ourselves hermeneutical.

Interpretation precedes conscious awareness of the perceived. However, in many respects, from a conscious perspective, Figal stresses that it is when we do not grasp or understand that we interpret, that is, try to understand, i.e., some complexity whose meaning escapes our attention yet in which we go on sorting out something making sense. Interpretation is dependent on an external world. Moreover, interpretation as such—in a world of givens—is unending. Unlike Figal, who views the pre-conscious flow of memory recontextualizations as a form of de-objectification, interpretation is, in contrast, an ongoing process of translation and transfer. In this process, past meanings and intentions are often obscured, acting as a force within the continuous flow of change and transformation, occasionally detaching from the objective.

[6] Figal (2010: 49).
[7] Ibid.: 50/52.
[8] George (2009).

The *role* of understanding is to address the increasing amount of interpretations due to the characteristic recontextualizing plasticity of interpretation itself. Understanding aims to establish a connection of what we grasp with the perceived object—of how effectively it aligns with its object. However, this does not explain what understanding is. What happens when we understand? We remember that Heidegger states that we are not *un-prejudiced* when we grapple with something we are trying to understand. *We use what we know in order to shed some light into the perceived opaqueness of that which we are confronting.* Understanding in *Being and Time* is an *enactment of the knowledge* of Dasein—and thus is *a self-understanding,* which is also *an understanding of the perceived.* Figal's complaint is that Heidegger narrows understanding to self-understanding.[9] It leaves out an aspect, which is essential to our everyday use of it—when our concern is not with *ourselves.* Heidegger made an important point, there is no understanding without enactment—understanding is a *realization.* To be able to enact an understanding an ability to do so is needed. Such an ability is an *acquirable* skill, according to Figal.[10] He exemplifies this by stating that "no one would say that one understands how to see or hear."[11] Besides the capacity to hear is the *skill to hear or see something in particular.* One has mastered something particularly well, something one is able to do *beyond* self-understanding.

The significance of the human realm is intricately connected to and reliant upon the realm beyond it—it embodies both the human "lifeworld" and the world of objects. The world comprises the interplay between the human domain and the domain of objects. Meaning emerges through our own actions and the external realm. Understanding, to differentiate it from pre-understanding, emerges from a state of not understanding and it always hinges on the ability to make intelligible what previously was unintelligible—and sometimes—in specialized circumstances—understanding hinges on the enaction of skilled ability. Not only is the skill an automated cognitive ability, what is understood, and through practice, will become familiar—and forgotten—and as such retrievable or enacted when the right context presents itself—to which it adapts by recontextualizing that particular automatized "forgotten" memory segment.

[9] Figal (2010: 90).
[10] Ibid.: 91.
[11] Ibid.

Then is the issue with perspectival perception and mediation. Nietzsche pointed to the "perspectival character of existence."[12] Everything appears differently when seen from different perspectives—every perception from a fixed perspective is a limited view onto the world, and which merges with the world. Remembering the previously mentioned *hermeneutic conversation* (Gadamer), or my own term *collaborative cooperation* in relation to medical clinical-technological interpretative practices, one can add perspectives, perceive of objects from various sides and angles, and commit the same matter to the scrutiny of different persons to complete a more thorough objective understanding of the matter. The more influences we allow ourselves, and the more different viewpoints we use to observe one thing, the more complete our understanding will be of that thing, the more objective it will be.

Freedom, language, and time are the three dimensions of Figal's theory of objectivity. Figal emphasizes space and hermeneutical space, which necessitates any formation of meaning; meaning that then is realized through freedom and language, and in what Figal terms the *time of enactment*.

Human activity turns on what is outside of it, in space, in the world it occupies. Hermeneutical space involves time in the order given in human temporal experience. Meaning is structured according to the temporal structures of experience. Figal is bringing in Husserl's temporal concepts of retention and protention.

Perceiving a specific pattern in, say, an X-ray image and subsequently reflecting on its informational content occur at separate instances in time. However, both elements are integral to the interpretation process and are not independent mental entities. Roman Ingarden, a Polish phenomenologist and a student of Husserl,[13] delineates how Husserl's concept of present perception, termed the "Here-and-Now" perspective, is divided into protentionality and retentionality. Retentionality denotes the innate capacity to maintain the presently perceived within the continuum of the lived present, sustaining its vitality, while protentionality anticipates forthcoming occurrences, drawing from past experiences to interpret outcomes. Consequently, the present moment is extended into a fusion of retention and protention, defining the experience of an event as unfolding presently.[14]

[12] Ibid.: 53.
[13] Roman Ingarden (1970).
[14] Ibid.: 143.

For instance, pronouncing "Constantinople" requires listeners to keep the initial phonemes alive while awaiting the completion of the word. Retention maintains the first part of the word, while protention anticipates the final part, thus experiencing the entire word in the present moment. This illustrates perception as actively constructing the phenomenon at hand, a pre-conscious act of apprehending and imposing coherence and meaning on unfolding events, which otherwise would be fragmented due to their incomplete nature.

When a radiologist examines MRI image sequences, similar retention and protention operate during the still-pre-conscious gaze at the image. Ingarden exemplifies this theory with personal experiences from a psychiatric ward in Krakow, where he observed fluctuations in a patient's emotions during a meeting between the two of them.[15] The girl's expression shifted abruptly from joy to horror, then to surprise and happiness, and back to horror and despair. Ingarden suggests that this experience illustrates the intimate connection between current perception and past experiences, where retention, perception, and protention merge.[16]

The enduring nature of perceived objects and their identity is closely linked to retention, without which identical objects cannot be constituted. In the case of the girl, it is proposed that her brain's ability to interpret visual perception may be compromised, resulting in a corrupted perception. Alternatively, this example suggests that, for most individuals, interpretation abilities precede conscious awareness in recognition processes. Figal comments that the Husserlian division of temporal experiences into protentionality and retentionality is showing the difference of primary and secondary memory—and that memory itself changes with the progression of experience.[17] Even the temporal character of perception is in itself retentional—"retentions of what has been perceived belong in perception (…) perception in an occurrence (…) is not the experience of something present, but, rather, of something coming and going. In the contemplation of a picture one can forget time."[18] The contemplation is itself in time—one looks at the whole picture, over there and over here, yet the picture is there all at once.[19] There is a perceptual openness to that

[15] Ibid.
[16] Ibid.: 148–151.
[17] Figal (2010: 274).
[18] Ibid.: 275.
[19] Ibid.

"something" occurring in perception. This invokes Heidegger's notion of *Gelassenheit zu den Dingen*—the "releasement towards things."[20] In our context this is a letting go of the preconceived, of one's attachment to theory, making oneself unfettered, receptive—and attentive "without a why."[21] It is not a total abandonment of oneself, instead it is an opening up to the meaning that crystalizes in this meeting of perception and image. The meaning cannot be *free* of human beings, instead the perceiver is initiated into the appearing meaning. In other words, we make ourselves receptive to what appears in experience, whose meaning is understanding. Openness is to detach from the noise of mind and hence to open the perceptually founded interpretative receptiveness necessary in order to "hear" the meaning. We make ourselves receptive to that which is other, which does not mean we lose ourselves. On the contrary, engagement and creativity appear simultaneously.

Detachment means turning our gaze away from ourselves and onto the externality, onto the other, which in Heidegger's thought is detaching ourselves from thinking as idea and will—and as Ruin writes, only by means of this change can the thing itself be understood.[22] This is nothing we can "will" to happen, or force through by thought itself. We have to be awake for it to change, be open for it to change, we must let it be. This releasement is beyond will, beyond activity and passivity, yet, as contrary this may seem, not beyond thinking and reason.[23] According to Ruin it designates a transformation of thinking itself, of how to think. It is not the kind of thinking we identify with representing and manipulation. It is a thinking that interprets, that is, openness to that which is other and hence presents itself. It involves "waiting" in which we open up to that which we await.[24] Thinking will hopefully reach a point in which it understands the other through contemplation of the situation and place in which it finds itself.

[20] Ruin (2019a, 2019b).

[21] See Caputo (1986), discussing the relation between Heidegger and Meister Eckhart, and Heidegger and the seventeenth-century German mystic Angelus Silesius.

[22] Ruin (2019a, 2019b: 5).

[23] Ibid.: 5–9.

[24] Ibid.: 6.

CHAPTER 8

# Temporality and Interpretation

We do perceptually frequent the things externally of ourselves, we enact the interpretation of the external situation. The structure of the enaction is temporal. Temporality is the structuring of experience. This is no different than the time structuring the occurrence of situations and both the evolvement and decay of things.

In other words, the temporal structure of our experience is subject to the time that governs the occurrence of things. The temporal structure is itself affected by situation, context, and occurrences we encounter—by the time governing the occurrence of things, which we correlate with. Helga Nowotny has written that "Time 'dwells' in us—through the biological rhythms to which we are subject, and because we are social beings who are born into a society with changing temporal structures and learn to live in its social time (…)."[1]

In contemporary neuroscience, scholars investigate the temporal concepts of retention and protention, with a particular focus on their implications within the domains of memory, perception, and temporal cognition.[2] These scientists aim at elucidating the mechanisms underlying the storage of past experiences and the preparation for future occurrences within the cerebral framework.[3]

[1] Nowotny (1994: 6).

[2] See, for instance, Grush (2006) and Nonaka and Yamaguchi (2022)—chapter 6.

[3] See Vogeley and Kupke (2007).



J. K. B. O. Friis, *Hermeneutics at the Intersection of Medical Technology*, https://doi.org/10.1007/978-3-031-88755-0_8

One prominent theoretical framework in neuroscience that addresses retention and protention is predictive coding[4]—which will be discussed later on. This theoretical perspective posits that the brain engages in the prediction of forthcoming sensory stimuli by utilizing prior experiential knowledge. Subsequently, these predictive models undergo continuous refinement and calibration in response to incoming sensory input. This iterative process entails the simultaneous retention of past experiences and protention of future events, as the brain dynamically adjusts its predictive expectations in accordance with newly acquired information.

Another example of an objective grounding of experienced time and temporality is Arthur Eddington's post General Theory of Relativity writing.[5] When Eddington writes about gaining insight into the physical condition that determines temporality, he suggests that the subjective mind can recognize an objective reality that is not merely an external condition. This objective condition must also be an internal property—a fundamental aspect of mind itself. In other words, it is a condition that belongs equally to human mind and body as it does to the rest of nature. Why should the human mind conceive of temporality in a form that is totally apart from the time of nature? Our conception of time, based upon experience, is as close as we get to a conceptualization of the physical condition, which supports experience[6]—as a condition for it. We lose any direct connection between the physical and mental aspects of the temporal experience, as we do not perceive the physical element itself but only experience it's unfolding. The physical aspect remains beyond our grasp—we do not sense it directly but infer its existence from the fact that our temporal experience is so fundamental that it cannot originate solely from our own perspective. As Eddington claims, we will always be able to recognize "becoming" because it is *insight*. It is insight because our elaborate nerve mechanisms do not intervene: "That which consciousness is reading off when it feels the passing moments lies just outside its door."[7]

We must simply come to terms with the idea that the mental insight into the time of physical nature is fundamental for any conceptualization we have about objective occurrences. That time also is a structuring of

[4] Clark (2015).

[5] Eddington (1946, 1976). Sklar (1995), also discusses Eddington's dual access to time.

[6] "Experience" encompasses a broader spectrum than merely serving as an "empirical method," as its foundation in *apprehending the world* extends beyond mere perception, observation, and measurement.

[7] Eddington (1946: 89).

mind, i.e., structuring the way we think and experience the world. The realism of the object depends on the mental insight into the flux that appears to us in experience. This experience also brings with it the realization of the significance of the experience; that, for example, we cannot reverse what appears. In this sense, we also have an *insight* into time's nature as "a kind of one-way texture involved fundamentally in the structure of nature."[8] We can know about this "texture" as we can also know about other properties of the external or physical world. We conceive of this transience as the passing of time, says Eddington, and furthermore, this is a "fairly correct appreciation of its actual nature."[9] We have one way in which we experience time directly. In order to "bridge the domains of experience belonging to the spiritual and physical sides of our nature," we need access to the world through our sense organs. We gain access to the temporal properties of external processes. We are able, through our sense organs, to relate time to other entities in the physical world. Eddington calls this "time's dual entry into our consciousness."[10]

We understand "becoming" because we, ourselves, *become*.[11] What is fundamentally involved in this expression of becoming? Clearly one aspect is that I have a body. In one respect, we are acting in a world that is constantly changing, and thus corresponding to the flux of sense experiences. In another respect, we realize that we are gradually becoming older. In activity, we produce entropy. I have—through my activity in the world I live in—brought time out in the open through my activity. This time is local and irreversible. That we produce entropy through our activity and, consequently, spend our energy is an objective measure for the transition of time. An objective time consists of local time in which the observers have dual access.

We experience the irreversible flow of events—a passage through which all other experiences are organized. This sense of time arises from our perception of changes in the external world: things appearing and disappearing, births and deaths, natural disasters, conversations, studies, our own aging, and the unfolding of our hopes and expectations—realized or not—through our actions. The time measured by the clock makes sense only because our experience of the world gives direction to the measure.

[8] Ibid.: 90.
[9] Ibid.
[10] Ibid.: 91.
[11] Ibid.

A hermeneutical realist provides us with interpretative accounts—or mediations—of things, objects, and occurrences because he also accesses the temporal ordering of the occurrences through the temporality embedded in the sense experiences of external phenomena of the world. Time's irreversibility must be explained both from the inside and from the outside. We begin with interpretation and experience and proceed to the concept and then to, say, the formalized symbol. However, even this course of experience is conditioned by our present awareness of the world and the continuous structuring and restructuring of memory. Only a subject that experiences things, objects, relations, and phenomena, all which are objects that are themselves under transformation by the fluctuation and dissipation of the world, can produce memory. As experiencing subjects, we are confronted with the visible traces of increasing order and disorder that are somehow *descriptive* for the flux presented to our senses. These traces are local changes. We experience these changes. They are transformations of the external world, of the phenomena that can be experienced, as well as of the observer as an organism.

No one denies the phenomenal reality of time—we undeniably perceive and experience events as unfolding within it. The crucial issue is that rejecting the reality of temporal experience would strip our way of perceiving the world of any ultimate significance, reducing it to mere illusion. To many theorists, it is self-evident that we exist in a reality where time is appearance. Others again see reality as part of the texture of wholeness, that is, which includes appearance among those phenomena of the world defined as real, and hence includes time, although in a rather inferior position. Still others reject the reality of time altogether because time, that is, a time conforming to the characteristics of experience, does not conform to the premises put down by "pure" thinking, premises which are beyond questioning. The answer to the question of the ultimate significance of time, which is beyond the reach of experience, is by this very fact beyond the reach of human cognition. However, evidently, *we are in time*. The time of mind that constitutes our awareness of presence *now* is fundamental to our experience. It is so fundamental that it cannot be separated from any kind of experience.

Perhaps the tendency is to reject any reality to temporality because we apply abstractions that cover up the transience itself. Abstraction enters every level no matter how one chooses to describe or explain the nature of time. Even in our everyday talk, which must refer to primitive, experienceable temporality, we have to apply idealizations and abstractions to a certain degree. When we refer to "now," or "yesterday," or "tomorrow," or "before" or "after," we are abstracting and applying the abstractions in our description of things and situations by creating a comprehensible and interpersonal order in our communication. We must understand that the human awareness of time is something evolving and that is a product of the interaction between humans and nature—it is an exemplification of how humans *adapt*. Temporality awakens through becoming; it awakens as experience, we encounter it continuously through perceiving. The meaning of the human world remains inescapably bound to and dependent on the external world.

Plant physiologists and chronobiologists[12] have extensively documented the intricate relationship and mutual influence between biological entities and their external milieu since at least the 1950s. From a biological standpoint, time manifests initially as an inherent component of metabolic processes propelling living organisms, progressively shaping various phenomena such as development, maturation, and evolutionary trajectories. Biological rhythms, encompassing daily and yearly cycles, synchronize harmoniously with external rhythms, establishing a profound connection between internal biological dynamics and the broader natural realm. Through understanding these correlations, biology acts as a conduit, bridging human temporal perception with natural occurrences.

By expanding upon this insight, we extend the discourse to examine the interplay between "organic time structure" and "phenomenological time structure," thereby arguing that organic factors influence phenomenological temporal experiences. It is important to note that this conceptualization does not entail a reductionist view of phenomenological time to biological time, a proposition deemed unattainable. Rather, the aim is to elucidate how the temporal framework of human experience is deeply rooted in the protracted evolutionary history of humankind, characterized by a progressively intricate interplay between biological processes and conscious cognitive faculties, adapting to and interacting with the environment.

Life thrives through dynamic processes of self-preservation and reproduction, each species creating distinctive systems. Reproduction, far from mere duplication, hinges on vital variations crucial for adaptation and survival. Driven by metabolic activities, these systems enable species to withstand and adapt to evolving environmental conditions, essential for

[12] Montaruli et al. (2021); Fishbein et al. (2021); Patke et al. (2020); Jagannath et al. (2017).

evolutionary processes and natural selection. Organisms maintain equilibrium with their surroundings while continuously undergoing transformation, underscoring the dynamic essence of life.

Life embodies a continual process of change and adaptation, where time serves as a cornerstone for evolution, maturation, reproduction, and ontogeny, delineating the essence of biological systems. Human temporal consciousness emerges through the interplay of phenotypes and genotypes, with adaptation being paramount for survival. Genetic components perpetuate through reproduction, fostering diversity and facilitating natural selection.

The passage of time influences both organic and inorganic domains, evident in natural rhythms, climatic variations, and organismal metamorphoses. Organisms exhibit diverse biological "clocks" to synchronize with external cycles, suggesting an early physiological grasp of time for survival. However, this understanding remains constrained by cellular biology limitations, necessitating reliance on genotype for adaptation.

Organic diversity reflects a spectrum of adaptations to temporality, particularly discernible in humans. These adaptations unveil sophisticated temporal mechanisms and life management strategies, underscoring the intricate interplay between life and the temporal fabric of existence.

Jerison's examination[13] of changes in brain-body size ratios indirectly underscores the fundamental role of time in survival, reproduction, and adaptation. Brain expansion likely occurred to augment sensory perception beyond vision, incorporating evolving senses like hearing and smell. This expanded pre-human perception necessitated novel ways of processing neural information, where both space and time emerged as critical dimensions for interpreting environmental stimuli.

Similar to linguistic principles, scholars such as Lestienne[14] and Campbell[15] posit the existence of regulatory genes governing adaptation to local environmental rhythms. This temporal adaptation involves a delicate interplay between innate biological rhythms and the external environment. Natural selection favors individuals adept at adapting to these rhythms through interactions between phenotype and genotype. Fraser[16] describes this interaction as "feedback," involving factors such as

[13] Jerison (1973, 1976).
[14] Lestienne (1995, 1998).
[15] Campbell (1982).
[16] Fraser (1990).

environmental dynamics and geographic isolation, shaping both phenotype and genotype evolution via mechanisms like gene flow and recombination, fostering novel gene combinations vital for adaptation.

Evolution thus unfolds as an ongoing process propelled by the intrinsic adaptability of living beings, perpetually preserving and augmenting coherence amid changing circumstances. Unlike the traditional notion of natural selection eliminating deleterious mutations, here, it is regarded as a mechanism for the emergence of advantageous traits. Nonetheless, time remains a fundamental attribute of the organic world, characterized by its ceaseless flux and transience, intricately weaving the fabric of existence.

Recognizing the presence of "natural rhythms" within metabolic systems underscores the symbiotic relationship between organisms and their surroundings. This acknowledgment distinguishes between short-term environmental influences, shaping immediate phenotypic responses, and long-term conditions, molding genotype adaptations over extended periods by observing environmental regularities and rhythms distinct from metabolic processes.

Organisms' metabolic systems, intricately linked with short-term and long-term aspects, influence fluctuating body temperatures. These variations affect an individual's time perception, altering the tempo of events. This internal interplay between metabolism and environment primarily shapes short-term responses, crucial for accurately gauging external temporal relations. While time comprehension is a product of human cognition, it necessitates synchronization between brain and body, incorporating both external and internal biological rhythms.

Rhythmic behaviors across all organisms imply an inherent connection to external and innate temporal cues, regulated by endogenous "clocks." These internal timekeepers, present everywhere across species, orchestrate behaviors independently of external stimuli. Notably, circadian rhythms finely tuned to day-night cycles and seasonal rhythms responsive to environmental shifts illustrate innate biological temporal mechanisms universally observed.

Research into flies like *Sarcophaga argyrostoma* and *Nasonia vitripennis*, including Saunders's studies,[17] has illuminated the complex mechanisms that regulate photoperiodic responses—changes in day length that prompt metabolic shifts such as diapause. Surprisingly, it is not the eyes, but rather brain receptors, that detect these light variations, emphasizing

[17] Saunders (1968, 1974, 1976, 2002).

the role of internal circadian oscillators in guiding behavioral adjustments. Saunders's findings reveal that flies have developed behavioral rhythms tightly aligned with environmental patterns.

Flies operate within two primary timekeeping systems, each with unique mechanisms adapted to specific species. These systems, known as the "interval timer" and the "Bünning/Pittendrigh" models,[18] showcase how internal rhythms synchronize with environmental cues to regulate biological processes. Bünning's model proposes that day and night are measured by an inherent daily rhythm divided into two half-cycles: one light-seeking (photophilic) and one dark-seeking (scotophilic), with dawn as the primary regulator of this rhythm. In this model, longer days extend the light phase into the scotophilic portion of the cycle, triggering responses characteristic of long-day conditions. Shorter days, on the other hand, prompt short-day responses. This timing mechanism is sustained by an internal circadian oscillator.

Building on Bünning's ideas, Pittendrigh[19] suggested that the "clock" is actually composed of two distinct oscillators, each one entrained separately by dusk and dawn. Saunders further developed this concept through his "internal coincidence model," which posits that photoperiodic changes adjust the phase relationship between these two oscillators. According to this model, light serves solely as an entrainment signal, activating the relevant oscillator. Extended daylight periods realign the oscillators, promoting continuous development without diapause, while shorter days cause a desynchronization that initiates diapause. Saunders's experiments lend strong support to this model.

The broader conclusion drawn from these models suggests that temporality in organisms like the flesh fly transcends simple hourglass or oscillator-based mechanisms. Rather, it represents a nuanced blend of both. This insight suggests that temporal awareness in organisms, including humans, is fundamentally biological but still deeply connected to the physical realm. Just as humans display behaviors tied to an inherent circadian system, they also share with flies a fundamental tether to physical reality, implying that biological time cannot be fully separated from physical time. As Harris aptly stated, "if there were no processes in the physical

[18] The Bünning/Pittendrigh model is a mathematical chronobiologic model describing circadian rhythms of plants and animals—developed by plant physiologist Erwin Bünning and biologist Colin Pittendrigh in 1970s and 1980s.

[19] Pittendrigh (1981).

world, there would be no one in the biological," underscoring the inextricable link between the two.[20]

Despite diminishing reliance on biological rhythms throughout human evolution, they persist in influencing vital physiological functions such as sleep cycles and menstrual periods. This underscores the foundational nature of human temporal experience rooted in biological processes. Our inherent physiological temporal framework, while capable of diverging from astronomical rhythms, remains deeply intertwined with external temporal cues, facilitating adaptive responses to the ever-changing environment.

Fraser,[21] Harris,[22] Jerison,[23] and Saunders[24] have collectively demonstrated the evolution of human cognitive faculties, including temporal perception, through processes of adaptation. Human temporality is deeply ingrained in nature's rhythms, and emerges through the synchronization of physiological rhythms with external environmental cycles. It is imperative to recognize that biological time seamlessly intertwines with physical time, highlighting the inseparable connection between temporal consciousness and natural phenomena.[25] This awareness, sculpted by physiological rhythms, fosters acute sensitivity to temporal dynamics and the relentless passage of time, intimately entwined with worldly events.

In other words, Figal's view on temporality—so important for the formation of meanings of a world apart from us, the other, the objective—is that temporality of mind is itself in time and as such in step with physical and biological analysis of time and activity. Expectation, attention, and memory, according to Theodore George, are never neutral—always embodied, situated, attuned, and always colored by attitudes and moods.[26]

[20] Harris (1988, 1993).
[21] Fraser (1990).
[22] Harris (1988, 1993).
[23] Jerison (1973, 1976).
[24] Saunders (1968, 1974, 1976, 2002).
[25] Harris (1988).
[26] George in his Introduction to his English translation of Figal's *Objectivity*.

CHAPTER 9

# Material Hermeneutics

Don Ihde's postphenomenological project expands Figal's hermeneutics of the objective, extending it into the realm of technologically mediated experience. This shift reimagines the role of objects in human experience, presenting them not as static entities to be passively observed but as dynamic presences actively shaped by the tools through which we engage with them. Central to this idea is the concept of "material hermeneutics," a term first coined by Peter Szondi[1] in 1995 and later developed by Ihde and Peter-Paul Verbeek.[2]

Through this lens, technologies are no longer seen as mere instruments for measurement or observation. Instead, they become active mediators of experience, shaping and transforming how meaning is generated. For instance, an Electrocardiogram (ECG) does not simply display raw data. It interprets electrical signals from the heart, detecting patterns such as an ST segment elevation, and presents them as visual readouts that medical professionals can perceive to diagnose conditions like acute myocardial infarction. In this process, the ECG itself becomes a co-interpreter of reality, embodying and enacting a hermeneutical engagement with the world.

However, this vision has not gone unchallenged. Within the broader field of continental philosophy of technology—and even within the

[1] Szondi (1995).

[2] Verbeek (2003).



J. K. B. O. Friis, *Hermeneutics at the Intersection of Medical Technology*, https://doi.org/10.1007/978-3-031-88755-0_9

postphenomenological tradition itself[3]—critical voices have raised questions about the nature and limits of technological mediation. These critiques have sparked vibrant debates, highlighting the evolving and contested nature of the field. The very tools designed to reveal the world are now scrutinized for their role in shaping it, underscoring the dynamic interplay between human experience, technology, and interpretation.

Materiality, according to Ihde, has two properties, such as being the target of scientific research and the application of technologies in this research.[4] In the postphenomenological approach there is the issue of *embodiment relations*. Such relations are something perceived or felt as extending the body of the perceiver by using the technology at hand, say, a *mirror* for parking the car or a blind person's *cane* gives the person a feeling of the free space behind or in front: **"I – technology – world**." The feeling of one body (I – technology) and my action is the arrow: **(I – technology) → world**. In radiology we use technologies to go beyond the bodily limitations of seeing—to see beyond what we normally can see by using technologies like X-ray or MRI to provide the image to view and the PACS to view the image. Here we are suddenly over in another kind of relation, namely, a *hermeneutical relation*: **I → (technology–world)**. Technologies like MRI mediate or "interpret" by applying magnetic fields, magnetic field gradients, and radio waves to generate images of the organs in the body, in other words, these are processes internal to the technology. The "I" (the radiologist) interacts with this mediation to generate a diagnosis based on his or her visual perception and interpretation or mediation of the content of the image. In addition to embodiment we have the interpretative aspect—the "I" mediates the technological mediation, or interprets the world as it is mediated by the technology.

Both in the PACS and the MRI modality there is an inbuilt "intentionality." This is an idea proposed by Ihde[5] and also Rosenberger and Verbeek.[6] This technological "intentionality" is, Ihde suggests, a built-in selectivity directing the radiologist how to use the PACS. Material technologies such as MRI and its images on the PACS direct the users to specific type of diagnostic-generating interpretations. Verbeek changes the

[3] See de Boer and Zwier (2024).
[4] Ihde (2009: 68).
[5] Ihde (2015).
[6] Rosenberger and Verbeek (2015).

relations by introducing an arrow of technological intentionality between the technology and the world: **I → (technology → world)**—however, in our case: **radiologist → (PACS → MRI → body)**. This is the scientific-technological *material* aspect of hermeneutics[7] where the technological mediations of the body are contained in the image.

Edmund Husserl's phenomenology provides an introspective analysis of experience, examining both the intending subject and the intended object. By bracketing assumptions, Husserl aimed to establish knowledge free from individual and social distortions, striving for maximum objectivity.

In contrast, Don Ihde's *Expanding Hermeneutics*[8] presents a distinctly different view of experience, one shaped by the century-long developments in social, technological, scientific, and philosophical spheres between Husserl's early twentieth-century work and Ihde's later contributions. Over this period, the phenomenological focus on introspection has been overtaken by a hermeneutic approach, American pragmatism, and the philosophy of technology within the continental tradition. Ihde contends that in the modern technological age, experience is no longer confined to individual subjective consciousness. Instead, technologies function as *extensions of human senses*, granting access to realities previously beyond human perception.

However, the information gathered through these technological mediations is far from "pure." Numerous challenges arise, both in how technologies detect input and in how observers interpret it. Technologies act as detection devices, collecting data—such as at nuclear energy levels—and translating it to a level perceptible by humans. Observers must then interpret and communicate this data, a process complicated by "observer variation," where different individuals perceive the same phenomenon differently. This is often due to human bias, undermining the accuracy of interpretation. Furthermore, technologies themselves are prone to error, producing artifacts that further disrupt the supposed purity of information.

Radiological research has addressed this issue by focusing on two primary strategies for reducing diagnostic errors. The first strategy relies on the belief that technological advancements can resolve observer-related problems by making abnormalities more visible. This includes optimizing

[7] Verbeek (2008).
[8] Ihde (1998).

viewing conditions, such as improving luminosity and monitor resolution, and enhancing diagnostic tools like artificial intelligence. While good technologies are essential, they are insufficient on their own.[9]

The second approach emphasizes understanding how radiologists interact with technological mediations, specifically how they engage with the information in images during the reading process.[10] Giving a twist to Manning, Gale, and Krupinski's assertion, we could say that the development, availability, and now usage of AI should not detract from the quality and need for radiological skills. One such crucial skill is visual perception.

## 9.1 Interpretation and Radiology as Object Mediating Technology

Image interpretation in radiology involves two closely connected cognitive processes: visual perception and the mental processing of that perception. While this framework forms the basis of radiological practice, it also presents notable challenges that merit closer examination. Radiologist Harold L. Kundel highlights that the core task of radiologists is image analysis, often guided by the assumption that their interpretations directly reflect the objective content of anatomical images. However, even with the use of advanced imaging technologies like PET and CT scans, which generate cross-sectional views of anatomy, radiologists must sift through an abundance of visual information. This process requires them to detect subtle abnormalities, often hidden within complex anatomical structures, underscoring the cognitive demands of accurate diagnosis.[11] Ultimately, the interpretation of medical images depends on both the human perceptual apparatus and the interpretive skills of the radiologist. However, Krupinski's use of the term "translation" seems to misrepresent the perceptual process; rather than creating a fully realized 3D mental image, the radiologist engages in pattern recognition within the 2D plane, identifying key structures and lesions based on experience and training.

Notably, at least 30% of all diagnostic errors in clinical practice are perceptual in nature. Consider the scenario of a radiologist tasked with identifying a lung nodule. Though the nodule's approximate location may be

[9] Manning et al. (2005: 683–685).
[10] Ibid.
[11] Ibid.

known, detecting it requires an extensive and systematic search process, which is shaped by the radiologist's expertise and understanding of where such nodules typically appear within the lungs.[12] Research involving eye-tracking technologies has revealed that search patterns are heavily influenced by clinical context, as well as the radiologist's professional experience.[13] This finding resonates with theoretical perspectives in phenomenology, hermeneutics, and the philosophy of perception, all of which emphasize the role of contextual knowledge, training, and experience in shaping human perception.

A poignant example of the influence of clinical history on perception occurred at a hospital in Copenhagen. Several years ago, an elderly woman underwent multiple surgeries and a series of X-rays after fracturing her femur. These images were stored in the hospital's PACS system. A year later, during a routine follow-up, eight new X-rays were taken—seven focused on the lesion and one encompassing her torso from the knees to the chest in profile. Another year passed before the woman was re-hospitalized with a large cancerous tumor in her stomach. During a subsequent morning conference, the radiologist responsible for her new follow-up reviewed the earlier X-rays. It became glaringly apparent to everyone in the room that the tumor was visible on the initial full torso image but had been overlooked. The error occurred because the radiologist had been solely focused on the area specified in the requisition. This type of oversight, known as "satisfaction of search," is a common cognitive error where the radiologist prematurely ends their search after identifying the first target, potentially missing additional, critical findings.

Radiological practice is inherently subject to observer variability, presenting a complex and enduring epistemological challenge. A notable aspect of this issue is the growing dependence on computer systems for the display of image data, which underscores the importance of ensuring that this information remains both perceptible and interpretable by radiologists. The specific imaging task at hand dictates the type of information required for accurate interpretation and diagnosis, thereby making image quality a pivotal factor. In this context, the effectiveness of radiological practice hinges on the seamless translation of image data into

[12] Krupinski et al. (1998: 611–612).
[13] Kundel (1990: 472–83).

diagnostically relevant information that can be reliably extracted by the observer. Kundel states that visualization and estimation tasks in image analysis rely on distinct visual cues to be performed effectively. Visualization tasks are primarily dependent on strong boundary cues to delineate structures clearly, while estimation tasks, such as gauging intensity, require a focus on texture cues. The effectiveness of both tasks can be enhanced by employing appropriate image displays. For example, the use of color can significantly improve the accuracy of intensity estimation, whereas three-dimensional displays offer substantial advantages in aiding visualization by providing spatial context. Ultimately, regardless of the nature of the task or the type of image being analyzed, the performance of the observer remains the definitive measure of the image's quality. It is the observer's ability to accurately extract and interpret the information that determines the utility of any given imaging modality.[14]

The interpretative efforts of radiologists are intricately tied to the technological tools they employ, which serve as essential mediators in the diagnostic process. As Don Ihde observes, these technologies facilitate radiological interpretation by making the process more efficient and accessible.[15] The devices used in radiology—MRI, CT, X-ray, and ultrasound—embody a "hermeneutic" quality, translating real-world phenomena into readable images.[16] Beyond merely converting data, these technologies enhance both the interpretability and objectivity of images by incorporating elements like color and graphical markers, refining the process of diagnosis. Ihde frames this technological mediation as a "material hermeneutic process," placing the interpretation of medical images at the intersection of the natural and human sciences.[17]

Imaging scientists, as noted by Kundel, aspire to predict how an observer will respond to any image configuration without conducting studies involving real human observers. This has led Kundel to advocate for "research into the deeper aspects of image perception and the interface between perception and analysis."[18] This call suggests a need for a deeper understanding of the hermeneutics of medical image perception.

[14] Ibid.
[15] Ihde (2009: 56).
[16] Ibid.
[17] Ibid.: 64.
[18] Kundel (2006: 402–408).

## 9.2 Medical Understanding

Hermeneutics is deeply embedded in the practice of medical image interpretation, as it is in all medical disciplines. Radiologists inherently employ hermeneutic principles when interpreting medical images, though they may not explicitly frame their process in these terms. For instance, they conceptualize it through "information-processing theory," a framework that underpins their approach to visual search and image interpretation. Despite this, the stages involved in radiological information processing closely mirror hermeneutics, as these involve a continuous interplay between observation, understanding, and contextualization of medically relevant phenomena.

In contrast to many other medical specialists, radiologists typically do not interact directly with patients. Instead, they specialize in the interpretation of medical images, focusing on specific patterns and leveraging their refined attention to detail and well-developed memory. As a technical and scientific discipline, radiology encompasses a range of specialized fields, each with expertise in different imaging modalities, including X-ray, computed tomography (CT), magnetic resonance angiography (MRA), positron emission tomography (PET), and magnetic resonance imaging (MRI). The interpretation of images from these modalities demands distinct cognitive approaches and interpretive frameworks, reflecting varied epistemologies and practices within the field. This diversity in interpretive methods challenges conventional views on scientific objectivity, as each modality offers a different "slice" of reality. Inevitably, interpretive errors can arise, posing challenges to the precision and reliability of diagnoses that hinge on human image interpretation. Perceptions drawn from imaging "data" are often incomplete, influenced by factors such as time constraints and cognitive load, which may further fragment the information received. Thus, radiologists frequently rely on shared interpretative frameworks to translate specialized insights into generalizable conclusions, creating a consensus accessible to non-specialists.

A typical day for a radiologist begins with the review of cases from patients admitted to the hospital or emergency department during the previous day and night. Their workspace includes two primary computer monitors and a third display known as the Radiology Information System (RIS). This system, positioned to the left of the radiologist, provides critical background information about each patient. The primary computer screen displays the images captured recently, while the secondary screen, positioned on the right, allows for comparison with prior scans of the same patient's condition. To start each case, the radiologist consults the RIS, familiarizing themselves with the patient's history, then examines the new images on the central screen, juxtaposing them with prior scans on the secondary screen for a comprehensive assessment.

This workstation setup exemplifies the complex cognitive and technical demands of radiology, as the radiologist synthesizes current and historical data, applying their specialized knowledge to form an interpretive understanding. Through such processes, radiologists navigate between the unique interpretative lenses provided by each imaging modality and an overarching clinical narrative, emphasizing the critical interpretive and analytical skills that define this specialized medical field.

Radiological interpretation often thrives on collaborative practices, where radiologists work alongside colleagues specializing in the same imaging modality. In these shared spaces—or sometimes in more isolated settings—radiologists encounter complex cases or ambiguous imaging data that invite collaborative inquiry. When interpretive uncertainty arises, radiologists consult their colleagues, posing questions like, "Do you see what I see?" Through this dialogue, they exchange insights, discuss observed patterns, and use additional technical tools and measurements to reach a consensus. Although multiple perspectives can reduce error, even shared agreement does not entirely eliminate the possibility of interpretive mistakes.

For the radiologist, interpretation is a deeply embodied and technologically extended practice, immersed within a social and interactive environment. As they adjust images with precise mouse movements, the radiologist's eyes scan for patterns while the brain, grounded in pre-understanding, aligns with the perceived details. Philosopher Jesse Prinz highlights the role of embodied pre-understanding, where prior knowledge contextualizes sensory input, allowing radiologists to respond to familiar patterns. Gallagher further emphasizes that V1 neurons, through predictive mechanisms, engage in both feature detection and anticipation based on past experience, effectively embedding priors within sensory processes that shape perception and responses.

The importance of interpretation in radiology cannot be understated, as it plays a crucial role in achieving accurate diagnoses and ensuring appropriate treatment. Nonetheless, errors still occur, often arising from a convergence of mistakes within the interpretative process itself. A considerable number of diagnostic errors stem from misinterpretations of radiological images, underscoring the challenges inherent in this field.

As philosopher Bas van Fraassen notes, "So here is the problem for epistemology: we take ourselves to have knowledge and to know what it is to be rational. Yet (...) in our past our presumed knowledge went into crisis (...). But what if we acknowledge that we could be in that position again?" This perspective reflects the persistent epistemic challenges in radiology, where interpretive errors reveal both the limitations of individual perception and the inherent uncertainties of knowledge.

Helmholtz famously described the limitations of the human eye's optics, noting that while the eyes are optimized for quick acquisition of broad visual information, they are less suited for detailed analysis.[19] In radiology, this limitation is further influenced by psychological and emotional factors that shape image perception. The interpretative process relies on mechanisms such as pattern recognition, spatial localization, and comparison with familiar templates—a cognitive approach known as Gestalt processing.[20]

When extracting details from complex image backgrounds, radiologists engage in two complementary processes that focus either on the visual field or the image itself. The first mode of perception is largely unconscious, automatic, and rapid, occurring through the brain's tacit engagement with the image. During this stage, the eyes perform saccadic movements, rapidly scanning the visual field to bring different areas into focus. This process is inherently fragmented and imprecise, as the eyes move at speeds of up to 400 degrees per second, producing blurred or noisy input. Remarkably, while the retina receives approximately 10 billion bits of visual information, only 6 million bits are transmitted through the optic nerves, and a mere 100 bits per second enter consciousness.[21] At this stage, the interpretation includes intuitive leaps, which can sometimes lead to erroneous conclusions, underscoring the complexity and potential fallibility of radiological interpretation.

The second mode of perception operates as a reflective, problem-solving approach characterized by slower, sequential, and deliberate processing that demands sustained cognitive effort. Reading an image within this framework unfolds across three performance levels. The first level,

[19] Helmholtz, quoted in Sabih et al. (2010: 1–2).
[20] Koontz and Gunderman (2008: 1156–1160).
[21] Sabih et al. (2010: 1–2).

termed "skill," entails cognitive patterns rooted in specific schemas or pre-programmed responses, shaping thought and action through habitual frameworks. Scholarly works by Sabih et al.[22] and Rasmussen and Jensen[23] identify the second level as "rule-based processing," wherein established rules facilitate responses to familiar problems through "if x, then y" patterns. The third level, "knowledge-based processing," employs analytic techniques to synthesize existing knowledge, enabling the inference of new insights.

The initial examination of an image typically yields a broad, global impression, in which the radiologist identifies fundamental elements—such as anatomical structures, color distributions, symmetry, and grayscale variations. This global impression draws upon long-term memory, shaping the radiologist's expectations through accumulated experiential knowledge. Although certain abnormalities may be apparent during this phase, the inherent complexity of medical images often requires a more targeted, focal search to discern subtle details. Pre-established mental frameworks or a skill-guided pre-understanding lead the way through this interactive process, where individual features are scrutinized and compared against prior experiences and relevant yet still tacit knowledge. A decision forms once these observed features align with the radiologist's understanding. The observer continually shifts between a global overview and focused search, synthesizing both holistic (the whole) and detailed perspectives (the parts).

Equally vital to the interpretive process is the effective communication of findings, which relies on the radiologist's structured and methodical approach. Clear, precise reporting of diagnostic insights not only ensures understanding but also strengthens *collaboration* among medical practitioners. Collaboration in radiology is essential for delivering high-quality patient care. A collaborative exchange between radiologists and other medical practitioners promotes more informed clinical decisions, as it allows for the integration of diverse perspectives and expertise. By working together, healthcare professionals can cross-validate findings, reduce diagnostic errors, and ensure that treatment plans are both accurate and comprehensive. This unified approach enhances patient outcomes by fostering a holistic understanding of the patient's condition, enabling more timely and targeted interventions. Furthermore, collaboration streamlines the

[22] Ibid.

[23] Rasmussen and Jensen (1974: 293–307).

diagnostic process, reducing delays in care and improving overall efficiency within healthcare systems. Ultimately, a collaborative approach in radiology not only strengthens diagnostic accuracy but also supports a patient-centered model of care.

In summary, in radiological practice human embodied cognition—the active process of interpreting—is intrinsically interwoven with technological mediation, creating a layered interaction with the world that precedes direct perception or imagery. Interpretation itself serves as a mediation, just as the radiological image perceived and analyzed by the viewer functions as a mediated representation of the original object of interest. For instance, when the radiologist examines an MRI on PACS, the image requires adjustments in terms of light, contrast, and other parameters. Technicians often further refine these images, tailoring aspects to align with the radiologist's preferences. Rather than distorting content, these adjustments reduce extraneous noise, sharpening the image to reveal the features most relevant to the diagnostic task. Ideally, this process yields an enhanced, clarified mediation for the radiologist, facilitating more accurate diagnoses and improving patient care.

The radiologist's interpretative and perceptual capacities are thus extended by technological means, which probe beneath the skin and render an internal view into a human-readable format. This technological augmentation extends the reach of hermeneutics.[24] Analogously, human embodiment and technological embodiment share this interpretive function, although in fundamentally distinct ways: the human body, for instance, interprets with emotions and sensory feedback, while algorithms process and render information in a disembodied form, devoid of affective influence. When humans interpret, they engage with the environment directly, while the radiological image remains a mediation of an aspect of the environment that cannot be perceived firsthand. The truth or validity of an interpretation or diagnosis depends on the skill of the interpreter, whose tacit knowledge and context-sensitive expertise draw on biologically rooted perceptual abilities. These interpretative acts are essential components of human cognition, which synthesize into the patterns that become diagnostically meaningful.

This cognitive foundation enables the detection of patterns that underpin diagnostic decisions, but as already mentioned, it is also subject to limitations. Interpretative errors may arise from internal factors, such as

[24] Ihde (1998: 44).

misrecognizing image patterns, which reflect both general and individual constraints in perception and understanding. This dynamic interplay, between human cognition and technological enhancement, highlights the cognitive complexities central to radiological practice.

## 9.3 Angioplasty as a Site of Collaborative Interpretation

The concept of medicine as interpretative praxis reflects the notion that clinical practice is not solely the technical application of scientific principles but also an art form requiring the nuanced interpretation of varied information sources to make informed clinical decisions. This perspective emphasizes that medicine is an interdisciplinary blend of scientific quantification and hermeneutic approaches, necessitating a synthesis of empirical knowledge with clinical acumen. Medicine, in this view, transcends rigid protocol adherence; it involves a dynamic process of interpretation, adaptation, and the integration of knowledge tailored to the unique values and needs of each patient.

Within clinical practice, interpretation encompasses the extraction of meaning from an array of technological outputs—such as medical imaging, lab results, and other diagnostic tools—as well as patient histories, presenting symptoms, and examination findings. This interpretive process is deeply rooted in a physician's expertise, accumulated experience, and understanding of medical science, while also being responsive to the individual characteristics and context of each patient.

Interpretation, a core skill in medicine, is inherently woven into the neural and biochemical networks that connect mind, body, and world. All healthcare professionals share this cognitive capacity, though its application varies widely across specialties. For instance, radiologists and cardiologists undergo similar foundational training, yet their daily interpretative practices diverge based on the specific demands of their fields. Radiology and cardiology thus represent distinct hermeneutic practices within medicine, each with unique interpretive focuses and methodologies.

In cardiology, interpretative praxis is applied directly in patient care, often within the immediate context of surgical or interventional procedures. Cardiologists rely on interpretative tools like electrocardiograms (ECGs), real-time monitors, and collaborative inputs from assisting colleagues in the operating room. Consider an on-call surgeon receiving a pre-dawn call about a man in his sixties with possible cardiac distress. The surgeon closely monitors ECG data streaming from the ambulance, assessing a series of readings for potential indicators of myocardial infarction. Recognizing the urgency suggested by the ECG, the surgeon arranges for air transport, and within twenty minutes, the patient arrives at the hospital's prepared catheterization lab for an emergency angioplasty to restore blood flow to the coronary arteries.

In this scenario, each interpretative step—from the analysis of the ECG to rapid logistical decisions—relies on the surgeon's capacity to synthesize real-time data with practical knowledge and tacit experience or skill, showcasing medicine as a dynamic and interpretative art in action.

Before the procedure begins, the patient is administered analgesics, sedatives, and anticoagulants as preventive measures. Once the patient is comfortably positioned on a cushioned table, the cardiologist inserts a catheter, typically through an artery in one of the arms or, in some cases, through the groin. Using live X-ray imaging for guidance, the catheter is navigated through the vascular system into the heart. A contrast dye is then injected, enhancing the visibility of blood flow on the monitor and helping the cardiologist pinpoint arterial blockages. Once located, a guidewire is threaded across the blockage, followed by a balloon that is inflated to reopen the vessel, restoring circulation. Ideally, the relief of the patient's chest pain signals the procedure's success, although, in certain cases, complications persist.

One such case involved a patient who continued to experience pronounced chest and arm pain after angioplasty. This unexpected development led the two attending cardiologists to review the imaging and reexamine the live feed. Unable to reach a conclusion, they consulted a senior cardiologist. Together, the team re-analyzed the feed and detected narrowing and constriction within the aorta. The cardiologist who had performed the angioplasty suggested a high-risk bypass surgery. However, the senior cardiologist proposed a less invasive alternative: inserting a stent. This procedure involved positioning a wire mesh tube, or stent, at the narrowed section of the artery, where it would expand with the help of a balloon catheter, restoring the artery's diameter and preventing future reclosure.

This case illustrates the profound cognitive and interpretive demands at the heart of interventional cardiology, particularly under conditions of uncertainty. When complications arise, cardiologists rely on collective expertise, examining images and live feeds, consulting colleagues, and deliberating on treatment options. These collective interpretive processes highlight the role of embodied cognition and the collaborative, communal nature of problem-solving in medicine, and of securing an objective insight into the problem at hand as well as a consensus for the solution.

The procedure exemplifies a complex hermeneutic process, as it requires dynamic interaction between brain, body, environment, and the interpreters present. In other words, this interaction encompasses social exchanges between the patient, nurses, and cardiologists, as well as constant monitoring of screens and tactile perception of the patient's arterial structures. The performing cardiologist interprets the tactile feedback from the catheter in conjunction with visual information from imaging, which then shapes the real-time adjustments made during the procedure. This dynamic interplay is not confined to the brain alone but reflects a broader embodied hermeneutic process that incorporates the operating environment and those within it. The cardiologist's hand movements and adjustments are continually attuned to the artery's physical response, while the brain responds to the evolving organism-environment interaction. Interpretation in this context is a process of dynamic adjustment,[25] where the brain, embedded within the larger bodily and environmental context, aligns with the task's demands through collective attunement and shared meaning-making.

When unexpected issues arise, the team engages in collaborative problem-solving, seeking senior expertise to interpret the findings communally. Interpretation in this setting is a collaborative cooperation in the operating room, where individuals jointly explore, communicate, and enact solutions. In other words, Collaborative cooperation unfolds through dynamic conversations[26] among team members, creating a live, interpretive interaction among clinicians, technicians, and the patient. Together, they engage in discussions informed by real-time medical data, such as the patient's heart and artery visuals, with the shared objective of discerning the best course of action for the patient's well-being. These interlocutors—medical professionals working across roles—actively listen and deliberate on issues that hold mutual significance, striving to reach a

[25] Gallagher and Allen (2018: 2642).

[26] Recall Figal's concept of "hermeneutic conversation," by "collaborative cooperation through conversation" is meant the same as Figal intends with his hermeneutic conversation, although the cooperation here involves interacting collaboratively with technologies, problem at hand, and information from the patient.

truthful understanding of the patient's condition. Here, "truth" is not assumed but emerges as a matter to be interrogated and clarified.[27] This provisional approach to truth, constrained to the immediate medical context, underscores a nuanced, situation-specific truth.

The interpretive process is intrinsically revelatory; through it, the participants often become aware of previously unexamined biases and potential misunderstandings surrounding proposed solutions. However, this self-awareness is incidental, arising as a byproduct rather than the primary aim of the discussion. Instead, the primary purpose is to understand the situation fully, which requires attentiveness to the complex web of relevant issues, judgments, and potential outcomes.[28]

In hermeneutic, conversational collaboration, truth itself may be contested. Different backgrounds, professional traditions, and personal experiences shape each participant's perception, leading to potential disagreements. Such variation is expected, as the issue at hand—multifaceted and intricate by nature—can be perceived through multiple valid perspectives.[29] Importantly, this conversation must remain open to the plurality of insights, as the most viable solutions often emerge from the diversity of perspectives brought into the interpretive exchange.

From another angle, in Gadamer's philosophy, the emergence of truth in understanding is conceived as an event or enactment of human existence, fundamentally a fulfillment of interpretative experience. Here, understanding is not merely intellectual but is woven into the very enactment of being. George[30] notes this approach but suggests, alongside Figal, that understanding transcends fulfillment as existence by emphasizing that it inherently involves a "reference" to the matter itself, independent of personal interpretation.[31] This enactment of understanding often occurs as a moment of insight—a sudden recognition of the issue at hand that provides the interlocutors with a shared point of concern in the interpretative process. Yet, this "insight" is not an absolute truth but an opening, inviting a collaborative exploration of the matter's potential meanings. The initial moment of understanding thus broadens into a search, where

[27] George (2022: 198).
[28] Ibid.: 199.
[29] Ibid.: 200.
[30] Ibid.
[31] Figal (2010: 66–77); George (2022: 201).

interlocutors deepen their experience by probing the interpretative possibilities, validating the initial insight through collective reflection.

Gadamer, however, does not fully investigate the trajectory of understanding beyond its initial enactment. He does not trace how understanding might evolve through continued engagement with the matter, nor does he explore the realization of the inherent possibilities of the subject itself. This limitation brings to light an essential aspect of interpretative practice: it must embrace a plurality of interpretations, as the matter at hand is rich with multiple perspectives and possibilities. This plurality is essential, not because of subjective divergence, but because the subject itself offers layers of meaning that cannot be entirely captured by a single interpretation.

An illustrative case is that of the cardiologists who, through their interpretative dialogue, collectively accepted the solution proposed by the senior cardiologist. This solution did not gain acceptance because of his seniority or any imposed authority. Rather, it became evident as the most effective and least invasive option—offering safety, reduced pain, and less complexity for the patient in contrast to a riskier open bypass procedure. The solution resonated with the other physicians not as a dogmatic assertion but as a pragmatic response to the immediate issue of the patient's arterial collapse. Gadamer would describe these physicians as "persons of experience," emphasizing that their interpretative engagement contributed to their *Bildung*, or self-formation. For Gadamer, this *Bildung* unfolds through what he terms the "dialectic of experience," which culminates in an openness to new interpretations and realities.[32]

This openness represents an intellectual and experiential readiness to engage with reality in its inherent ambiguity. Reality itself, Gadamer suggests, is complex and often resists definitive understanding; thus, any engagement with it demands flexibility and a willingness to reinterpret. It is through these dialogical encounters that the interlocutors not only deepen their understanding but also cultivate a refined sensitivity to the limits and potentialities of human knowledge in the face of reality's inexhaustible complexity.

[32] Gadamer (2013: 364); George (2022).

CHAPTER 10

# Forensic Interpretation

Forensic sciences, originating from nineteenth-century efforts to support law enforcement through medical expertise, have evolved into a robust field grounded in a rich tradition of interpretative practices. The discipline now incorporates an array of advanced techniques and technologies—sophisticated imaging modalities, cutting-edge analytical methods, and comprehensive statistical models—pushing the field to transcend biases rooted in subjective experience. Despite these technological strides, forensic practitioners continue to grapple with fundamental challenges surrounding the reliability and validity of interpretative methodologies.

Debates within forensic science often center on epistemological issues, particularly regarding the interpretation of visual data. Imaging, photography, radiology, and fingerprinting are all practices heavily reliant on visual perception and analysis of complex data, including traces, images, and biological materials. At the heart of these debates lies the critical question of how visual evidence should be understood and the role that human perception and interpretation play in this process, especially as they relate to imaging techniques.

Hermeneutic issues arise when investigators and scientists make interpretative decisions, whether framing a crime scene or analyzing the nuanced data collected. These interpretative choices are inevitably shaped by the cultural and experiential backgrounds of the practitioners themselves, underscoring the integral role of human perception in forensic science.



J. K. B. O. Friis, *Hermeneutics at the Intersection of Medical Technology*, https://doi.org/10.1007/978-3-031-88755-0_10

## 10.1 Forensic and Hermeneutics

Interpretation is foundational to all forensic sciences, yet the term "interpretation" itself demands a clear and precise definition. Many non-forensic fields, including medicine, law, physics, and mathematics, also involve interpretive processes, often within the domain of "hermeneutics." In forensic contexts, addressing interpretation as a form of knowledge generation, particularly through a hermeneutic lens, is essential for understanding how evidence is analyzed and understood.

The emerging field of forensic hermeneutics merges hermeneutic philosophy with forensic analysis, applying interpretive methodologies to the examination of evidence and artifacts within legal and investigative settings. This interdisciplinary approach enables forensic scientists to uncover deeper meanings, intentions, and contexts that may not be immediately apparent in the evidence alone.[1] One significant hermeneutic method used in forensic science is the Ricoeurian framework of "narrative identity analysis," which recognizes that understanding our social world often relies on collectively negotiated narratives that give shape to complex human interactions. Another approach builds on the philosophies of Heidegger and Gadamer, emphasizing the role of pre-existing knowledge and implicit understanding. Heidegger posits that interpretation is not a mere labeling of objective phenomena; rather, it reveals the inherent relevance of events, people, and contexts as understood through our own prior knowledge and experience.

Forensic hermeneutics thus positions interpretation as a vital part of forensic practice, aimed at dissecting and comprehending various forms of evidence, including documents, witness statements, crime scenes, and digital information. The goal is not just to establish facts but to uncover layers of meaning, intent, and interconnections within the evidence. Importantly, forensic hermeneutics acknowledges that interpretation is never entirely objective; it is shaped by the interpreter's background, biases, and assumptions. Consequently, a hermeneutic approach to forensic analysis involves careful examination of one's preconceptions and tacit knowledge, which influence interpretive decisions.

Forensic hermeneutics has practical applications across numerous areas. For example, it aids in examining written documents such as contracts, letters, or diaries to determine authorship, authenticity, and intent. In

[1] Sonsino (2005); Ovens (2013).

criminal investigations, forensic hermeneutics enables the analysis of language patterns in speech and text messages to assess suspects' motives or understand contextual aspects of crimes. Additionally, hermeneutic methods play an essential role in cybercrime investigations, where digital evidence like metadata, timestamps, and communication logs must be interpreted to trace cyber activity and establish timelines.

Moreover, forensic hermeneutics extends to historical and archaeological inquiries, where artifacts, art, cultural objects, and historical texts are examined to authenticate origins, uncover hidden meanings, and interpret symbols. This interpretive framework allows for the reconstruction of past events, supporting investigations into historical crimes or uncovering narratives embedded in cultural artifacts.

As an evolving discipline, forensic hermeneutics seeks to bring structure and methodological rigor to the interpretive processes in forensic and legal investigations. By emphasizing critical analysis, interdisciplinary collaboration, and reflective examination of biases, forensic hermeneutics contributes to the accuracy and credibility of evidence interpretation, fostering a more robust foundation for decision-making within forensic science.

## 10.2 Forensic Imaging

Forensic imaging involves the application of visual documentation methods to capture and preserve evidence critical to criminal investigations. Although it holds a significant role within forensic science, the practice has not been without controversy. A primary concern relates to the potential for inaccuracies and interpretative biases, which are issues shared by other medical fields, such as radiology. Observer variability, a common challenge in data interpretation, is especially pertinent in forensic imaging, where different examiners may interpret identical images in divergent ways. This variation has raised concerns about the reliability and validity of findings.

Within forensic radiography and radiology, however, there is a prevailing belief that technological advancements can help reduce the subjectivity inherent in human interpretation. Haldar suggests that technologies like X-ray imaging have transcended "hermeneutic practices" and asserts that such tools minimize human agency, creating a form of knowledge that is "purely objective."[2] This perspective sharply contrasts with other

[2] Haldar (2013: 130 and 142).

areas of radiology, which, as Haldar discusses extensively, remain deeply enmeshed in collaborative interpretation.

Outside of forensic radiography, other forensic imaging modalities—such as photography, microscopic analysis, digitized images, and DNA analysis—still heavily rely on human interpretation, which inherently brings in bias and observer variability.[3] In recognition of these concerns, the U.S. Department of Justice revised its testimony guidelines in 2018, disallowing unequivocal claims from forensic units like the FBI's Forensic Audio, Video, and Image Analysis Unit regarding pattern matching, acknowledging the potential for bias and error in such analyses.

In discussions among forensic radiologists, the accuracy of postmortem imaging remains a focal point.[4] Forensic radiography, while valuable in crime scene investigations, has encountered criticisms regarding the absence of standardized protocols and guidelines.[5] Unlike medical radiography, which operates under rigorously established standards, forensic radiography lacks universally accepted protocols.[6] This gap in standardization contributes to inconsistencies in procedural approaches, image acquisition, interpretation, and reporting. Such variability complicates the comparison and corroboration of findings across different forensic radiography examinations, ultimately challenging the consistency and reliability of forensic radiographic practices.

Forensic radiography, particularly in postmortem examinations, involves numerous ethical complexities that necessitate careful consideration.[7] Respecting the deceased's dignity and privacy remains paramount throughout the imaging process, and care must be taken to ensure that these values are upheld during virtual autopsies or postmortem CT scans. The advent of these minimally invasive imaging techniques, while offering alternatives to traditional autopsies, prompts ethical inquiries about the balance between invasive and non-invasive procedures, especially in cases where the results may directly impact legal decisions and criminal investigations.

A further ethical dimension lies in the interpretive nature of radiographic analysis. Forensic image interpretation requires expert judgment,

[3] Gabrielson (2019).
[4] Beck (2011); Eriksson et al. (2017); Kavitha et al. (2009); Aalders et al. (2017).
[5] Aalders et al. (2017).
[6] Eriksson et al. (2017).
[7] Ibid.; Beck (2011).

and the subjective nature of this interpretation can lead to varying conclusions among radiologists and forensic pathologists. These discrepancies may arise from differences in individual expertise, variations in training, or the quality of imaging technology, all of which can affect the accuracy and reliability of forensic conclusions.[8] Such interpretative subjectivity can bear significant consequences for criminal investigations and judicial outcomes, underscoring the need for a critical examination of the role of subjective interpretation in forensic radiography and its potential effects on the objectivity of forensic science.

Concerns raised by theoreticians within forensic science underscore significant questions about the interpretative processes employed by experts. There is a discernible shift toward favoring inductive reasoning and statistical methodologies. Rather than navigating multiple, often divergent interpretations, the trend favors more streamlined, reductive procedures aimed at enhancing consistency and reliability. This methodological preference reflects an effort to construct a framework that prioritizes reproducibility and minimizes the subjective variability traditionally associated with forensic analysis. Hermeneutics would, on the other hand, have emphasized the importance of interpretation in expert analyses, and questioned the recent trend toward reductive methodologies like inductive reasoning and statistical models. While these methods seek to build a consistent and reproducible framework, they risk oversimplifying complex cases and overlooking the nuanced insights that interpretation offers. Interpretative processes, with their capacity to consider multiple perspectives and accommodate contextual variation, can enrich forensic analysis by embracing a range of potential outcomes rather than reducing them to a singular result. In emphasizing the value of interpretive expertise, forensic science can develop a *collaborative* framework that both respects intersubjective reliability and acknowledges the depth that interpretative flexibility brings to nuanced investigations. However, we need to look closer at the forensics in practice.

## 10.3 Interpreting Visuals

Courts often depend on experts across diverse fields to reach informed and just verdicts. Experts, with specialized training in scientific or technical domains, contribute valuable insights but are also expected to embrace

[8] Eriksson et al. (2017).

the interpretative dimensions of their work, rather than simply reducing uncertainties to definitive answers. Forensic radiologists, for instance, interpret complex radiological images where nuances are critical, and benefit significantly from collaborative interpretative exchanges with colleagues. This collaborative interpretation embraces the complexities inherent in image analysis, ultimately enhancing the reliability of expert testimony in legal proceedings.

A prominent debate within forensic science centers on the role of visual perception and interpretation in generating meaningful insights. Forensic practices frequently rely on the analysis of visual data—such as images—where interpretation and subjective judgment are essential components. These interpretive processes, however, as we all are well aware of now, are inherently vulnerable to biases that shape our understanding. David Hume poignantly addressed this issue in 1777, noting that our limited knowledge of causal relationships often leads us to rely on past experiences, thus forming beliefs based on what we already know while overlooking our "ignorance" of unexperienced possibilities.[9] This interpretive tendency influences both experts and laypeople, though experts, due to their extensive training and specialized experience, are more adept at minimizing error. Nonetheless, embracing interpretative expertise rather than striving solely for eliminative and reductive objectivity can deepen our understanding by allowing for nuanced, context-sensitive insights that are critical in forensic science.

The effectiveness of human visual perception is closely tied to its adaptability for specific purposes. Unlike the vision of certain animals, such as birds, which is highly specialized, human vision is designed for flexibility. This adaptability equips humans with specialized mechanisms that allow for rapid responses to unexpected stimuli or movements. Yet, the visual system operates largely below the level of conscious awareness, leading individuals to sometimes perceive things inaccurately—either seeing what is not there or overlooking what is.

As previously discussed, the brain engages in a pre-conscious process when interpreting visual information, drawing on memory to construct and fill in details. This process occurs before the perception reaches full conscious awareness, allowing the brain to create a seamless and coherent interpretation of the visual input. This understanding, however, is inherently approximate; conscious perception lags slightly behind real-time

[9] Hume (1990: 56).

events, as the brain must create plausible interpretations that keep pace with the "speed" of the unfolding reality. This interpretive approach means that what we ultimately "see" is not a direct sensory experience but a refined and plausible interpretation shaped by both experience and intuition.

From an interpretive standpoint, human perception is inextricably shaped by individual experiences, which may not always align directly with the visual stimuli being sensed. Although extensive training and repeated exposure can help reduce these subjective influences, personal biases in perception can never be entirely eradicated. Some degree of observer variability is inevitable.

As David Hume posits, cognition is not the product of abstract, detached reasoning but rather arises naturally from associative processes, habits, passions, and perceptions. In essence, our thoughts and judgments are interpretations rooted in habitual tendencies built from past experiences. When humans interpret visual information, they engage implicitly in inductive reasoning, drawing conclusions based on patterns observed over time. The key challenge, therefore, is not to "overcome" or surpass current perception, but to recognize that inductive reasoning is an intrinsic aspect of human interpretation. Embracing this understanding allows us to better appreciate the interpretative nature of perception.

## 10.4 Interpretation and Individualization

When an individual contacts a surface with the palmar side of their hand, they inevitably leave behind a fingerprint impression on that surface. This impression comprises unique patterns created by the natural ridges and furrows on the skin of the fingers, organized in random but identifiable configurations. As documented by Virtual Labs[10] and widely accepted within forensic science, these patterns are inherently unique to each individual, forming a personal signature that enables specific identification. In forensic investigations, fingerprints discovered at a crime scene are meticulously documented, often through forensic photography, and serve as essential evidence. By tracing these fingerprints back to their source—a process known as individualization—investigators follow a core forensic protocol standard in criminal investigations.

[10] Virtual Labs (2023).

This analysis raises intriguing questions regarding the role of perceptual interpretation within forensic science. How has the concept, based on the psychological notion that past events often predict future outcomes, come to underpin forensic methodologies? Furthermore, why has inductive reasoning, central to perceptual interpretative processes, become foundational in forensic science practices? These questions invite a closer examination of the ways psychological principles and reasoning methods contribute to and enhance forensic analysis.

The forensic understanding of fingerprints as unique to each individual is deeply rooted in historical scientific methodology. This foundation can be traced back to Galileo (1564–1642), whose principles for scientific inference emphasized the use of mathematics and the experimental method to advance knowledge through observation, induction, and generalization. Inductive reasoning and interpretative methods allow observers to perceive signs as indicators of phenomena extending beyond direct observation. Early applications of these techniques emerged in medicine, particularly in fields like autopsy and toxicology, which the legal system soon incorporated to inform criminal cases.[11]

By the eighteenth and nineteenth centuries, conjecture and interpretative reasoning began to supplement the Galilean scientific approach in medical diagnostics, leading to innovative identification methods. The principle of individualization draws from the work of Gottfried Wilhelm Leibniz (1646–1716), whose "identity of indiscernibles" posited that entities indistinguishable from each other must be identical.[12] Lambert Adolphe Jacques Quételet (1796–1874)[13] expanded on this, suggesting that nature inherently avoids repetition. This idea later influenced Alphonse Bertillon (1853–1914), who proposed that individual measurements could serve as perceptual markers of similarity. Bertillon introduced the "product rule" for determining the joint probability of independent events by multiplying their separate probabilities, leading to a classification system to identify repeat offenders.[14] His system relied on measurable anthropometric features, such as height, weight, head circumference, and limb measurements, and he incorporated probability from physics to account for measurement errors, applying these methods to criminology and social science.

[11] Hald (2014).

[12] Forrest (2010); Saks and Koehler (2019).

[13] Saks and Koehler (2019).

[14] Hald (2014).

In contrast, Francis Galton (1822–1911) posited that each person carries visually discernible, unique characteristics, even in their smallest features, asserting that there is a "design" within each individual that contributes to species diversity. Galton provided probabilistic evidence for personal identification using fingerprints, establishing their uniqueness through statistical analysis. Thus, the primary theoretical challenge in forensic science lies not in observation alone but in how inferences drawn from observation are validated and informed, underscoring the centrality of inductive reasoning to the discipline.

It is significant that David Hume (1711–1776), in the 1777 posthumous edition of *An Enquiry Concerning Human Understanding*, critically examined the rationale behind inductive reasoning.[15] Hume argued that induction—the practice of forming universal principles based on specific observations—lacks a rational foundation. He questioned the assumption that the future must mirror the past, noting that even consistent patterns in experience do not logically necessitate such uniformity. For Hume, our reliance on nature's consistency and future predictability stems from mental habit rather than rational justification. He famously contended that if we relinquish the belief in the rationality of induction, we must acknowledge that it is "not reasoning which engages us to suppose the past resembling the future, and to expect similar effects from causes which are, to appearance, similar."[16]

This tension regarding the rational basis of induction is central to the forensic sciences' discourse on uniqueness and individualization. In forensics, the need for reliable identification is paramount: law enforcement, forensic experts, and the judicial system rely on the capacity to link evidence to an individual with certainty to achieve justice and administer appropriate consequences.[17] However, this process hinges fundamentally on the concept of personal identity—a concept whose justification remains, as Hume observed, philosophically tenuous under traditional inductive reasoning frameworks.

The debate over inductive reasoning in forensic science involves two primary "fallacies" that are frequently employed to comment on those who rely on this method for forensic conclusions. The first is the "uniqueness fallacy," a concept identified by David Balding,[18] which emerges when

[15] Hume (1777 (1990), sect. IV, Part II).
[16] Ibid.: 39.
[17] Leone (2021).
[18] Balding (2015).

legal professionals—attorneys, judges, and DNA experts—presume that a set of genetic markers, expected to appear in fewer than one in five billion people, must necessarily be unique. The error here lies in deducing uniqueness solely from low-frequency occurrences, without sufficient consideration of the broader context of available samples. Contemporary forensic science underscores that physical evidence, even when attributed to a specific source, does not guarantee a singular, definitive identification of that source.[19]

The second, more foundational fallacy is the "individualization fallacy." This fallacy stems from the assumption that it is feasible to achieve absolute individualization, a process by which an object or trace is categorized into a class of one, thus implying its uniqueness.[20] The notion of individualization has fueled debates for over thirty years, with statisticians and proponents of inductive reasoning defending its scientific validity, while critics argue that inductive methods lack the rigor required for conclusive forensic application.

Supporters of absolute individualization often justify their position through a combination of metaphysical reasoning, appeals to authority, and a form of "wishful thinking." The metaphysical argument holds that no two objects are identical, as they will inevitably reveal observable differences upon closer examination.[21] Appeals to authority rely on the assertions of prominent figures within forensic science, both historical and contemporary, to reinforce the plausibility of individualization. The "wishful thinking"[22] argument posits that inherent variability among objects will always allow for differentiation, based on the assumption that unique identifying characteristics are always present. However, scholars Saks and Koehler contend that these arguments rest on a foundation of belief and intuition rather than empirical evidence. The uniqueness hypothesis thus assumes that each person has inherently distinct physical characteristics—such as fingerprints, footprints, or handwriting—that can serve as exclusive identifiers. This hypothesis, however, remains a subject of contention, particularly as forensic science evolves and reassesses the limitations of individualization in practice.

[19] Ibid.; Saks and Koehler (2019: 205).
[20] Stoney (1991: 197).
[21] Saks and Koehler (2019: 209).
[22] Ibid.

Critics of the individualization concept—including scholars such as Saks and Koehler, Balding, Stoney, Mingxiao Du, and Morrison[23]—challenge the premise that each object or biological trace is inherently unique and argue against the necessity of definitively identifying a single source for evidence found at crime scenes. Balding, for instance, notes the practical impossibility of confirming the distinctiveness of any human characteristic across all individuals, observing that without testing every individual, uniqueness remains unproven.[24]

Supporters of the individualization hypothesis maintain that scientific rigor is achieved through advanced technological methods and precise quantitative analyses. Jamieson,[25] for example, argues that while technological measurements may carry the risk of error, they still provide concrete data points, such as drug composition, tool mark impressions, or the weight of evidence samples. Yet, the interpretation of these measurements requires more than raw data; it involves hermeneutic processes in which evidence is given meaning through contextual analysis. The effectiveness of this interpretive approach relies heavily on the specific context and the expertise of the forensic interpreter. However, it is not without challenges, as cognitive biases and interpretive errors can significantly impact the final conclusions drawn from forensic evidence.

Additionally, there is a marked deficiency in cross-disciplinary collaboration within forensic science and law enforcement, which poses risks to both accuracy and understanding in criminal investigations. Forensic inquiries draw upon a range of specialized disciplines—each employing distinct data types and interpretive frameworks. When forensic experts from diverse fields work in isolation, their data interpretations are later synthesized to suggest possible individualizations of a suspect, yet without the benefit of interdisciplinary collaborative dialogue, inconsistencies and interpretive errors may arise.

The absence of collaborative exchange among these specialized fields can lead to significant risks. Misinterpretations of technical data may go uncorrected, and errors may be compounded by a lack of contextual understanding. This issue becomes particularly critical when evidence from one discipline strongly implicates a specific individual, while data

[23] Saks and Koehler (2005, 2019); Balding (2015); Stoney (1991); Du (2017); Morrison (2022).

[24] Balding (2015: 54).

[25] Jamieson (2008).

from another does not corroborate this finding. Without effective interdisciplinary collaboration, such conflicts in data interpretation may result in biased or incomplete conclusions, undermining the reliability of forensic evidence and potentially leading to erroneous outcomes in the identification of perpetrators. The integration of perspectives across fields is therefore essential for a *balanced*, robust *approach* to forensic analysis, ensuring that evidence is thoroughly interpreted within a holistic framework.

The individualization debate underscores the complex intersection of scientific methodology and subjective interpretation in forensic science, highlighting the potential limitations of relying on individualization as a conclusive identification method.

Jamieson contends that inductive reasoning is fundamental to forensic science, relying on consistent system performance to validate hypotheses.[26] This approach aligns with the Bayesian framework, which combines prior beliefs with new evidence to assess hypotheses in light of updated probabilities. While this method may lack a purely logical foundation, it is widely deemed necessary, given the probabilistic nature of scientific evidence and the challenges of achieving certainty in complex cases. Bayesian inference allows forensic scientists to systematically update the probability of a hypothesis as new evidence emerges, integrating prior knowledge with observed data. This probabilistic reasoning is particularly valuable since absolute certainty is often unattainable due to the inherent variability in evidence, such as fingerprint analysis, or trace material comparisons.

Contrasting with Jamieson's view, which emphasizes scientific rigor through the technological processing of data, Mingxiao Du[27] raises significant concerns about the reliability of quantitative and technological methods in forensic science. Du points out that the integration of technology and instrumentation in data collection opens the door to various errors. Every measurement is inherently accompanied by uncertainty, and even with advanced instruments, absolute accuracy remains unattainable. Differences between measured values and actual values result in what Du categorizes as systematic, random, and gross errors.[28]

Systematic errors are, like in radiology, attributed to factors such as the methodologies applied, the instruments used, environmental conditions,

[26] Ibid.
[27] Du (2017).
[28] Ibid.

and human factors. For instance, slight temperature variations during instrument use can yield inconsistent measurements across different operators. Proper calibration can mitigate systematic errors, but a failure to do so constitutes a procedural error that may compromise findings. Conversely, random errors are unpredictable and arise from factors beyond control, which undermines measurement reliability. Du emphasizes that these random errors scatter around a central value in a pattern that is irregular and unpredictable, complicating precise interpretation and highlighting the limitations of human perception. Such errors reinforce the notion that measurements are approximations rather than absolute values.

Gross errors, arising from human error, typically occur when incorrect values are recorded during instrument operation. Du argues that perceptual and interpretive challenges extend beyond data collection to affect how forensic experts interpret traces as unique evidence of a particular perpetrator. These interpretive hurdles reveal that human perception and technological limitations together shape the forensic process, suggesting that measurements should be regarded as approximations rather than definitive truths. This perspective calls for a cautious approach to asserting uniqueness in forensic evidence, as the reliance on technology and human interpretation remains prone to inherent uncertainty.

## 10.5 Forensics and Memory

The contemporary approach to assessing forensic sciences increasingly integrates insights from psychology and the cognitive sciences.[29] Initially centered on understanding cognitive biases, forensic cognitive science research has since broadened to encompass diverse psychological and cognitive elements within the forensic domain. This expansion now includes the study of human perception, memory, contextual information, and experiential factors in forensic practices.

Unlike fields such as medicine, where cognitive science has long played a crucial role, forensic science has only recently begun to incorporate these insights. A growing body of documented errors within forensic disciplines has underscored the necessity of psychological and cognitive expertise to better navigate the complexities of cognitive fallibilities that can arise during forensic analysis.

[29] Edmond et al. (2017: 145).

Consider, for example, the before-mentioned intricate process of fingerprint comparison. Examiners are often tasked with comparing two unfamiliar and distinct fingerprint impressions side by side to determine if they originate from the same or different fingers.[30] While one might assume that this process relies exclusively on deliberate and objective perception—unaffected by memory or past experiences—research reveals that perception is indeed shaped by an examiner's prior experiences.

Experience in this context can have both advantageous and detrimental effects. For instance, through extensive training and experience, latent fingerprint examiners develop the ability to interpret and differentiate between most fingerprints at a glance, even when visual noise is present or under time constraints. This expertise allows experienced examiners to achieve significantly greater accuracy than novices. However, prior experience can also introduce biases. When a current fingerprint case shares similarities with past cases, such as visually similar fingerprints or analogous case details, these resemblances may inadvertently influence judgments in image comparisons. Additionally, research indicates that fingerprint examiners' judgments may lack internal consistency over time and external agreement with other examiners. Such findings demonstrate that perception in forensic analysis is not purely a product of the visual information available but results from a complex interaction between that information and external, often unrelated, cognitive factors.

A particularly contentious practice within forensic science is the reliance on eyewitness testimony, which underscores the complex and often unreliable nature of human memory. Memory is susceptible to errors, yet these lapses often occur without our conscious awareness of them. Cognitive scientists advocate that forensic professionals prioritize documentation and empirical data over memory due to its fallibility. However, memory and interpretation are inherently intertwined in cognitive processes, with interpretation drawing on memories. Understanding the characteristics and limitations of memory is thus crucial in forensic contexts, as studies show that 72% of wrongful convictions are linked to errors in eyewitness memory.[31]

Visual perception further illustrates the depth of neural interpretation involved, with memory being actively restructured to create contextual

[30] Ibid.

[31] https://wmich.edu/sociology/causes-wrongful-conviction

references that shape understanding even before conscious awareness. Human interaction with the environment is largely automatic and effortless, distinct from the mental exertion required in tasks like problem-solving or textual analysis. This automaticity of perception, as already mentioned, is influenced by social and cultural contexts, as meanings become internalized over time. These meanings continuously resurface through a feedback cycle involving pre-conscious perception, neural interpretation, and memory recall. Within this semi-conscious state, humans often react to incomplete sensory information, forming conclusions based on partially processed stimuli.

The limitations of visual perception are particularly pronounced when it comes to interpreting scientific data. Data only becomes "evidence" through processes of selection and modification, such as enhancing contrast or resolution in an electronic X-ray image. This transformation process brings the central challenge of interpretation to the forefront: raw visual information must be processed and framed within a context, with biases and prior knowledge shaping how forensic data is ultimately understood and presented as evidence.

Historically, eyewitness testimony has been a cornerstone of criminal investigations, providing firsthand accounts of objects, individuals, and locations pertinent to the crime. As direct observers, eyewitnesses can recount details based on personal observation and are often called upon to identify suspects through lineups or to describe their appearance for use in composite production systems that generate likenesses of potential perpetrators.[32]

However, research by Loftus,[33] and Loftus and Laney,[34] has demonstrated that eyewitness identifications frequently lead to mismatches. This inconsistency is largely due to inherent cognitive limitations in how people encode and recall facial features, even in cases where they recognize the individual. Studies reveal that facial recognition is generally unreliable and susceptible to biases, while the methods used to assist eyewitnesses in identifying suspects often fail to account for these limitations.[35]

[32] Hodge (2018).
[33] Loftus (1992).
[34] Laney and Loftus (2008).
[35] Green (2018).

The Innocence Project underscores the critique of individualization—that forensic testimony can also inadvertently overstate the significance of perceived similarities between crime scene evidence and suspects, sometimes oversimplifying the complexities involved in the identification process. Such overemphasis can mislead courts, compromising the objectivity that forensic science seeks to uphold. Together, these findings point to the need for critical scrutiny of eyewitness evidence, advocating for approaches that mitigate bias and acknowledge the cognitive challenges inherent in human memory and perception.

CHAPTER 11

# Embodied, Embedded, Experienced, Extended

The experience of perceiving objects in the world is undeniable and fundamentally grounded in our engagement with our surroundings. For instance, when I consider Berkeley's evocative example of "kicking a stone," the resultant pain is not a detached intellectual concept but rather a direct consequence of my physical interaction with the stone. Such perception encompasses not only clear and distinct objects but also ambiguous scenarios, like interpreting partial shadows as a half-face on a wall. Even highly skilled observers may occasionally perceive things that lack an external material basis. These perceived yet non-existent entities are mind-dependent, as they emerge from the brain—and body. What we experience as "seen" is often a recontextualized interpretation of sensory input. When we cannot fully comprehend what we are observing, our minds frequently default to familiar forms, effectively filling in gaps with recognizable patterns. This perceptual process operates beyond our conscious awareness, managed by neural mechanisms that interpret and structure our sensory world without our explicit control.

In cognitive science, a division exists between advocates of direct realism, who argue that we have unmediated access to the world as it truly is, and constructivists, who maintain that perception and observation are deeply influenced by implicit frameworks and pre-reflexive intuitions, which shape and assign meaning to our experiences. However, the boundary between direct realism and constructivism is not as distinct as it may



J. K. B. O. Friis, *Hermeneutics at the Intersection of Medical Technology*, https://doi.org/10.1007/978-3-031-88755-0_11

initially appear. Constructivist approaches, such as Predictive Processing (PP),[1] posit that our knowledge of the perceptible world is actively constructed, yet this does not imply an inability to obtain reliable knowledge. Instead, distinguishing between direct and indirect perceptual content proves challenging, as perception often involves a dynamic interplay of immediate sensory data and the brain's interpretative processes. Notably, research indicates that when perceptual information is incomplete or ambiguous, the brain compensates by interpolating details to support coherent and actionable perceptions. To appreciate the interpretative role of bodily-perceptual engagement with the material world, it is essential to examine the nuanced distinction between direct and indirect access. This inquiry is particularly relevant in contexts like radiology, where perceptual inaccuracies occasionally occur and underscore the complexities inherent in interpreting visual data.

## 11.1 Perception and the Role of Interpretive Mediation[2]

Our perceptions are continuously subject to distortions of reality, often without our awareness. Don Ihde[3] highlights that our perceptual limitations become evident through our reliance on technology to access realities beyond our direct experience. Immediate perception offers access only to isolated fragments of the broader reality, reminding us that much of the world remains beyond human perception. Three primary influences shape our perceptual experience: personal specialization, cultural context, and biological foundations.

Individualized perceptual skills develop within social and cultural frameworks. Expertise in areas such as music, visual arts, and diagnostic imaging demonstrates the impact of specialized training on perception. Through rigorous testing, we refine our interpretations, assessing their alignment with objective data. Nonetheless, it is crucial to acknowledge that these interpretations are inherently human. When engaging with technologies like nuclear magnetic resonance imaging (NMRI), we are working with mediations that capture and represent energy forms far beyond our natural perceptual abilities.

[1] Clark (2015: 12–15).

[2] This chapter is a variation over the problem of perception and interpretation discussed in a previous Commentary in *Found. Sci* (2012)17:363–367.

[3] Ihde (2011).

Our biological nature, rooted in evolution, lies largely outside our immediate cognitive awareness. Evolution has equipped our brains with mechanisms for survival-oriented perception, fostering a pre-conscious ability to process fragmented sensory information over time. Both nature and culture, however, exert conflicting influences on our perceptual capacities, highlighting an epistemological challenge inherent in Ihde's concept of "translational mediation." This challenge lies in interpreting an image's gestalt within a context that transcends human sensory experience—particularly when the biological-neurological processes underlying interpretation are beyond direct human control. The result is often a tension within our bodily-perceptual roles, leading to interpretative ambiguity.

The philosophical exploration of perceptual interpretation has deep historical roots. Democritus, one of the early thinkers on this subject, distinguished between direct sensory perceptions and the mind's contributions to perceptual experience. This distinction may have stemmed from a metaphysical pursuit of "truth-to-(human)-nature" or personal experiences of sensory conflict, such as feeling warmth while others feel cold. John Locke's well-known "bucket experiment" also underscores this conflict: after immersing one hand in warm water and the other in cold water, placing both hands in lukewarm water produces simultaneous sensations of warmth and cold. For Democritus, such secondary sensory qualities—warmth, coldness, and colors—were not intrinsic properties of objects but mental constructions within the observer's mind. This insight continues to resonate, as it underscores the mind's role in constructing sensory experience even when technology extends our perceptual reach.

An intriguing concept emerges when we consider the mind's role in "filling in" certain sensory details to complete experiences of warmth, coldness, or their combinations. This process suggests that we actively contribute qualitative aspects to our sensory experiences that exceed the objective data presented to our senses. Locke, influenced by Robert Boyle's studies in sensory physiology, explored this differentiation between primary sensory qualities (e.g., pressure, weight, vibrations) and secondary qualities that we subjectively experience—a distinction inspired by Democritus and Descartes. Locke aimed to make this difference accessible through straightforward experiments, which demonstrated that perception consists of several stages: (a) the object observed, (b) the light refracted from it and detected by the eye, (c) physiological processing, and (d) the resulting mental image or idea. Locke's work established a lasting dichotomy in perception theories: the objective reality of the external object versus the image perceived by the observer.

Building on Locke's foundations, Hermann von Helmholtz advanced the understanding of perception, emphasizing that it is not merely a passive process but involves active construction. His insights laid the groundwork for modern constructivist theories, in which perception is seen as an interpretative process that operates through "translational mediation," a concept later explored by Don Ihde. Helmholtz described this as an "unconscious conclusion," a term that has since permeated contemporary perception research. Terminology varies among scholars, with some preferring "unconscious inference" while others use "filling-in"; although there may be subtle distinctions, all these terms describe an underlying mechanism that imparts meaning to sensory input.

The core distinction between Locke's and Helmholtz's perspectives lies in Helmholtz's recognition of perception as a complex, multi-layered process. This process involves far more than the inward projection of a retinal image; it is a synthesis of sensory data, pre-existing expectations, and fragments of knowledge, all of which converge to form an integrated perception of the world. Helmholtz argued that this amalgamation produces perceptual thought, or "unconscious inference," which actively shapes our experience of reality.

As Yantis suggests, perceptual processing "gives rise to veridical experiences despite the limitations and distortions of our senses."[4] Yet, the object perceived is never experienced in its objective purity; rather, it is imbued with elements derived from the unconscious inference mechanisms—or predictive processes—that shape perception. Helmholtz explains that these "mixed" objects appear within the field of vision as if they would produce a consistent impression on the nervous system under ordinary, stable conditions.[5]

The cultivation of expertise and the refinement of perceptual skills significantly shape our capacity to interpret technologically mediated images. At the core of this process lies the subjective nature of translational mediation, wherein we encounter an image or "mediation of a specific energy level." While we perceive and interpret this image through our senses, it is in this decoding process that tacit elements of unconscious inference

[4] Yantis (2001: 21).
[5] Ibid.: 21 & 25.

become active, embedding themselves as implicit scaffolding for perception and subtly shaping the content of what we perceive.

It is crucial to recognize that these tacit elements consist of fragments of information that may bear little direct relation to the object of interpretation. These elements operate beyond conscious awareness, presenting a "hermeneutical horizon" that can significantly challenge the pursuit of "veridical experiences." These personalized "mind-elements" consistently influence perception, subtly imparting interpretive meaning to images produced through technologies such as X-rays, ultrasounds, and NMRI scans.

The value of any interpretation lies in its effectiveness for the specific purpose it serves. Human vision, while versatile relative to the vision of birds or other animals, is equipped with specialized mechanisms—such as the rapid redirection of gaze toward unexpected motion or instinctive reactions to environmental stimuli. Marr[6] suggests that the unique functions fulfilled by the visual systems of various species require tailored representations, resulting in distinct perceptual realities. Thus, human vision has evolved to address the particular needs of our species. Crucially, however, we remain largely unaware of the elements that have shaped this customization of our visual system; in other words, we do not yet possess a complete "picture" of how our vision is constructed. As a result, we may either perceive what is not actually there or fail to notice what is.

The human brain interprets perceptual objects and unfolding events instinctively, pre-consciously, and in an approximate manner. In effect, the brain "fills in" perceptual gaps before they reach conscious awareness, drawing on a personal reservoir of memories, instincts, and experience. Concurrently, the brain constructs an "imaginary temporal loop" around the event in real time, helping to approximate the continuity of experience even as conscious perception lags slightly behind objective reality. This process enables us to keep pace with the rapid flow of events by constructing simplified, plausible models of what we perceive, essentially completing the perception for us.[7]

In other words, our perception is supplemented with additional content that can either enhance accuracy or introduce biases. Although extensive training and repeated exposure to specific tasks—such as medical imaging interpretation—can reduce these issues, personal influences on

[6] Marr (2001: 118).

[7] Dennett and Kinsbourne (1992:183–247).

perception can never be entirely eliminated. Consequently, a degree of variation among observers will always persist.

In medicine, the epistemological challenge posed by differing interpretations of the same image—termed "inter-observer variation"—is a critical concern. Imaging technologies, including X-rays, ultrasounds, and NMRI, are indispensable in modern healthcare, enhancing clinicians' perceptual reach well beyond the inherent limitations of human senses. While algorithms and computational tools can sometimes be employed to "read" or "decode" images for diagnostic purposes, human interpretation remains essential and is often preferred due to its nuanced adaptability. However, interpretations grounded in individual perception inevitably reflect subjective predispositions and personal backgrounds, which shape each clinician's approach to image analysis.

Inter-observer variation reflects the degree of discrepancy that can arise between multiple observers when interpreting the same image, even when those observers share comparable training, experience, and social backgrounds—as they also constitute the foundation for collaborative interpretations; however, without such cooperation the differences may contribute to mistakes. What accounts for these perceptual differences? As Marr has noted, while humans are biologically standardized to an extent, only certain elements of human perception may be universally consistent. Tacit knowledge, memory, and lived experience play a pivotal role in shaping individual perception. These perceptual frameworks are influenced by diverse sources, including personal history, cultural background, and social context, all of which can vary significantly—not only across cultures but also within the same society, often along social or professional lines.

For instance, individuals with specialized experience, such as members of organized crime syndicates, highly trained Special Forces soldiers, or humint field operatives, demonstrate heightened situational awareness and sharper perceptual acuity compared to those without such backgrounds. This disparity highlights the profound influence of prolonged experience, environmental conditioning, and social context on perception.

In fields like medicine, where precision and accuracy are paramount, recognizing these perceptual differences is crucial. Radiologists, for example, rely on their accumulated experience to detect subtle anomalies in medical images. Understanding how tacit memory shapes perception underscores the importance of training, experience, and context-awareness in improving diagnostic accuracy and minimizing errors.

Pascal Mercier articulates a compelling vision of human perception as a flawed and often almost comically inadequate attempt to convey the depths of an internal world that remains largely hidden. This inner world, with its complexity, continually strives to express itself outwardly yet rarely achieves full articulation. Mercier suggests that these underlying structures of perception are not merely a function of biological instinct but are also deeply intertwined with cultural and social origins.[8] In essence, perception emerges from a reservoir of inherent biological responses that initially guide actions, but these responses are further shaped and refined by cultural learning and experience. Thus, perception is not purely a physiological process; it is a blend of brain-based neural structures and the acquired insights of cultural context.

Experimental psychologists Hannah Faye Chua, Julie E. Boland, and Richard E. Nisbett[9] have shown how cultural influences play a substantial role in predictive processing or shaping our unconscious inferences. Their research explores how individuals build mental models of perceived stimuli within working memory. They emphasize that these interpretations are not faithful replicas of external reality but are often simplified and lack fine detail.[10] The phenomenon of memory, too, is inherently subjective and colored by one's present circumstances. Recollections are filtered (restructured) through the lens of current experience, and memories emerge as stories partially shaped by ongoing life events. This selective nature of memory implies that while certain aspects of past events are highlighted in the present, others become obscured, gradually receding into the "neuronal labyrinth" as J.L. Carr poetically observes in *A Month in the Country*. Over time, the past becomes a faint and altered landscape, existing somewhere between memory and the present moment.

Chua, Boland, and Nisbett's study provides an empirical basis for understanding how culture shapes perception. They analyzed the visual behavior of two groups: Asians born and raised in the United States and Asians born and raised in Asia, who shared biology but differed in cultural upbringing. When presented with complex images, their eye-tracking results revealed notable cultural differences in gaze patterns. Asians raised in the United States tended to focus primarily on the central object within an image, with little attention to the surrounding context. Conversely,

[8] Mercier (2007: 24).
[9] Chua et al. (2005).
[10] Enns (2004).

Asians raised in Asia distributed their gaze more broadly, attending to both the central object and the background context. This research underscores that our habitual use of vision and directed attention is a learned behavior influenced by the environment in which we are raised and socialized.

However, philosopher Don Ihde offers a contrasting view, emphasizing learned perceptual skills and expertise while seemingly downplaying the role of biological influences. Ihde proposes that the interpretative process depends on decoding visual information as if it were structured and complete. But how reliable is this assumption, given that sensory input may be inherently fragmented? This raises a critical question: if perception is, by nature, an act of interpretation, can we fully trust our "translational mediations" to deliver objective reality, or are we always working with a constructed version of the world around us?

In essence, the gestalt of perception is shaped by training, social influences, and a biologically evolved nervous system predisposed toward certain interpretative leaps. The challenge is determining the relevance of the constructed information we bring to perception. Can we ever achieve a "truth-to-nature" interpretation through perception alone, or are we continually bound by an imperfect blend of innate and learned influences that shape how we interpret reality? These questions underscore the complexity of understanding perception and hint at the limits of achieving an absolute correspondence between perceptions and so-called objective truth.

## 11.2 Direct or Indirect Access to the World of Objects and Things

As *hermeneutists*, we encounter inherent limitations in our ability to probe the neural and biochemical foundations that underpin interpretation and understanding. The specific cultural, historical, traditional, and social influences that shape our interpretive acts often evade clear delineation. While we might observe certain preferences or inclinations shaping our interpretations, these observations raise profound questions: What precisely constitutes the foundational background that informs an interpreter? What types of "data" exist within what we call "effective historical consciousness"? And how are these elements selected and recognized as significant?

Gadamer's perspective offers a valuable insight here; he asserts that interpretation is not entirely within our volition but primarily an occurrence beyond conscious intention and control. This view may invite hermeneutists to explore the origins of perceptual gestalts—the foundational structures of tacit understanding and the source of meaning. However, despite this invitation, hermeneutists are constrained by a lack of methodological tools suited to such inquiries. The investigation of interpretive processes encounters a barrier at the edge of consciousness, as phenomenal experience does not permit us to traverse beyond-consciousness realms. Consequently, we are precluded from experiencing phenomena in their original, unmediated states.

Despite these constraints, hermeneutics offers other promising avenues for exploration. These alternative pathways extend beyond the limitations imposed by text, language, and conscious awareness, deriving their potential from within the hermeneutic tradition itself. Rather than focusing solely on what is explicitly articulated in language or text, these avenues encourage an exploration of the pre-structured backgrounds that inform understanding—those implicit interpretive frameworks that, while inaccessible to direct consciousness, still shape interpretive acts.

Through these pathways, hermeneutists can probe the historically and socially constituted backgrounds of interpretation, not by examining them at a neurological level but by analyzing the structures of tradition, discourse, and intersubjective experience that shape them. These frameworks encourage us to consider interpretation as both a personal and collective inheritance—a dynamic interplay of influences that are, as Gadamer suggests, "effective" in the sense that they act upon us and shape understanding, often without explicit awareness.

While hermeneutics may not allow for empirical access to the neural or subconscious origins of interpretation, it does offer methods for understanding the historically conditioned and socially constructed elements that guide interpretive acts. The task, then, becomes one of uncovering the implicit frameworks within our interpretive backgrounds and understanding how these tacit influences shape the phenomena we interpret. This approach acknowledges the limits of consciousness while remaining committed to the hermeneutic project of understanding meaning as a dynamic, culturally and historically situated process.

## 11.3 Hermeneutics and Brain

The intersection of neuroscience and hermeneutics might appear to be an unlikely pairing, yet computational neuroscientist Péter Érdi and mathematician-neuroscientist Ichiro Tsuda have proposed a conceptual bridge between these fields. In a series of papers, both collaboratively and individually, they have explored the brain's intrinsic hermeneutic processes—an area of study that has intrigued neuroscientists since the 1980s.[11]

In the hands of Érdi and Tsuda, hermeneutics gains new significance within the neuroscientific domain. Their research suggests that the brain functions as an interpreter of sensory information, constructing a layered and intricate web of meaning from the raw data of perception. Érdi succinctly encapsulates this view by asserting, "The brain is itself an interpreter: the brain is a hermeneutic device."[12] This perspective does not adhere strictly to classical philosophical frameworks of monism (viewing the brain as a singular, indivisible substance) or dualism (which posits a separation between mind and body). Rather, it promotes a more nuanced, integrated approach to understanding cognition and perception, positioning the brain as an active participant in the construction of meaning.

Their work, spanning from the 1980s through the early 2000s, has influenced subsequent research on cognitive processes, highlighting the interpretative capacities of the brain as central to the human experience. By conceptualizing the brain as a hermeneutic entity, Érdi and Tsuda laid the groundwork for an interdisciplinary dialogue that merges insights from neuroscience, philosophy, and cognitive science to better understand human cognition and perception.

During the 1980s, it became increasingly necessary to compare human cognitive processes to those of artificial intelligence (AI). At that time, AI systems demonstrated an ability to process finer details in visual data than the human eye, underscoring some of the limitations inherent in human perception and cognition. AI's method of processing relies heavily on probabilistic calculations, aiming to model objectivity by minimizing subjective interpretation. In contrast, human cognition is fundamentally interpretative, which, while limiting the potential for purely objective perception, also enables a unique, contextually rich understanding of the world. This interpretative nature led many researchers to conceive of the

[11] Tsuda (1984, 1991, 1993). Érdi (1983, 1996, 1999). Érdi and Tsuda (2002).
[12] Érdi (1996: 179).

brain primarily as a computational device, aiming to model cognitive processes in terms of algorithmic outputs and suppressing the subjective, intuitive aspects of human understanding.

Ichiro Tsuda's research in neuroscience embraces a dynamic view of human perception and cognition, resonating with principles from Japanese Zen Buddhism and the earlier theories of autopoiesis advanced by Maturana and Varela. Published four years after Maturana and Varela's work, Tsuda's contributions provide a neuroscientific perspective on hermeneutic processes within the brain, emphasizing how interpretive dynamics shape perception and cognition.

Central to Tsuda's approach is our previously discussed concept of *Vorverständnis*, or "pre-understanding," which underpins the brain's interpretive processes. Pre-understanding, according to Tsuda, functions as a tacit, pre-conscious framework informing immediate actions and decisions, especially in situations requiring rapid responses. Unlike tasks that allow time for reflective thought, such as interpreting a complex text, immediate decisions rely on an internalized foundation of past experiences and acquired knowledge that operates largely below conscious awareness. When engaged in interpretive tasks, however, the individual can consciously explore a text or problem in relation to this background knowledge, which, while not entirely present in consciousness, deeply informs the interpretive process. This movement, classically described as the "hermeneutic circle," does not bring us back to the same interpretive standpoint from which we began; rather, understanding deepens and evolves as one engages with the material. Tsuda suggests that the metaphor of a "spiral" more accurately reflects this interpretive movement, as our comprehension grows and transforms with sustained engagement.[13] This interpretive spiral enables a departure from purely logical processes, which derive conclusions strictly from premises, and instead allows meaning to emerge as an intuitive, recursive interaction between consciousness and pre-understanding.

For Tsuda, interpretation depends on a feedback mechanism between sensory input and the subconscious framework of pre-understanding. Sensory data first engages the subconscious, where it undergoes initial processing, then transitions to conscious awareness, where meaning is constructed. Tsuda describes this as a "feedback circuit," in which sensory input is modified by subconscious processes before becoming accessible to

[13] Tsuda (1984).

conscious interpretation. Meaning, according to Tsuda, cannot be logically deduced; it arises only through interpretation, as the brain continually contextualizes new information in light of existing pre-understanding. This interpretive feedback involves a sequence where each new part of an experience updates the pre-understanding, which in turn reshapes the next interpretive phase.[14]

Tsuda also posits the existence of "hidden dynamics with hidden variables" in the brain's interpretative process. These hidden variables encompass a wide array of influences internalized over an individual's lifetime, encoded within neural memory networks. As such, they become accessible and active when triggered by relevant stimuli, though they remain largely opaque to conscious awareness. This hidden dynamics operates on what Tsuda calls the "meta-conscious" level, allowing the brain to process multiple streams of information simultaneously, though often at the expense of detailed accuracy. The pre-understanding, comprising traces of past cognitive processes, provides a subconscious scaffold for interpretation. During an interpretive act, the visual cortex and other relevant areas integrate this pre-existing structure with new sensory data to inform perception. Here, Tsuda emphasizes that each interpretative moment is a fusion of past and present, with no clear break between the two; understanding always arises as a synthesis of historical cognition and immediate perception. Through this fusion, Tsuda's work highlights the brain's role not as a computational device but as a dynamic, interpretive entity capable of constructing meaning through continuous, evolving interaction with its own experiential history.

## 11.4 Bayesian Neuro-hermeneutics

The concept of "chaos" has often been invoked to describe the underlying mechanisms that drive the cerebral cortex's dynamic generation of meaning.[15] According to Walter Freeman, the brain not only processes sensory inputs but actively generates meaning, transforming raw input into a meaningful construct that provides information about the environment. Freeman suggests that reality may sometimes be identified as a mental construct,[16] raising questions about the nature of this construct: does it

[14] Ibid.

[15] Freeman (1991: 78–85).

[16] Ibid.

distort reality, or does it faithfully represent it? Does the idea of a mental construct inherently imply a distortion?

In Freeman's view, the oscillatory process of "pre-understanding -> reality -> pre-understanding" does not necessarily distort reality. Rather, pre-understanding—the cognitive background shaped by previous experiences—serves as an interpretative lens grounded in real-world interactions. This foundational structure, or "pre-understanding," is enabling the brain to interpret new inputs meaningfully without necessarily altering the essence of those inputs.

The question of what constitutes a "correct" interpretation remains complex. A correct interpretation may not be strictly objective, yet it remains valid if it aligns with experiential coherence and functional utility within a given context. Here, accuracy does not imply a mirror-like reflection of external reality but rather an interpretation that integrates sensory data and pre-understanding in a way that is both contextually accurate and meaningful to the individual. Thus, the construct of reality that emerges from this process is not a mere distortion but a deeply informed, dynamically generated understanding rooted in both present stimuli and past learning.

Karl Friston and Chris Frith propose that active inference—a process they describe as probabilistic and composed of discrete, singular operations—offers a framework applicable to hermeneutics, standing in contrast to Ichiro Tsuda's view that interpretative processes are deterministic and extend beyond mere inference. Tsuda argues that deterministic derivations are fundamentally distinct from interpretive processes; however, Friston and Frith, in contrast, contend that the challenge of Tsuda's hermeneutics can indeed be addressed through active inference.[17]

According to Friston and Frith, interpretation functions as an inference derived from a specific perspective. Such interpretation or inference is, therefore, inherently inductive: it forms conclusions based on observed individual instances or real-world events.[18] Predictions emerge from these inductively inferred conclusions. For example, if one observes someone eating a cookie and grimacing, they might infer that the person dislikes the cookie. Friston and Frith's approach further incorporates a metaphysical dimension into this model by introducing the concept of "representation," based on a Bayesian framework of cognition. Here, per-

[17] Friston and Frith (2015: 129–143).

[18] Ibid.: 129.

ceptions are seen as interpretative or inferential processes grounded in the capacity to predict based on sensory stimuli. According to Bayesian theory, these predictions are shaped by neural representations formed through prior experiences.

The notion of "representational" cognition assumes that higher-order logical or analytic thought retains experiences in memory in a way that preserves the original "imprint" of these experiences over time. This preserved imprint allows the brain to recognize and respond to new, similar events by referencing stored experiences. Such a process is known as inductive reasoning, paralleling the rational approach that philosopher Dagfinn Føllesdal suggested as an alternative to hermeneutics through hypothetical-deductive derivations. In this view, rationality underpins interpretative processes, allowing them to function as inductive, experience-based operations that align with Bayesian probability models rather than purely deterministic or non-inferential modes of understanding.

Interpretations are not exclusively products of inductive reasoning. Sensory receptors are continuously engaged in collecting environmental input, which is then interpreted by the brain to inform actions and perceptions. This interpretative process in perception involves both *bottom-up* and *top-down* processing. In bottom-up processing, sensory receptors detect external signals, which the brain then integrates and processes into coherent information about the environment. Conversely, top-down processing is shaped by prior experiences, knowledge, skills, and memories, which dynamically interact with new sensory inputs. These memories are actively structured in relation to the current context, continuously updated through a feedback mechanism. This feedback illustrates the hermeneutic cycle of "pre-understanding—part—whole," highlighting interpretation as a dynamic process in which understanding evolves as new aspects emerge.

In contrast, the Bayesian approach widely adopted in cognitive science conceptualizes the interpretative process instead as statistical reasoning. Here, the brain functions by calculating and updating probabilities, converting raw sensory data into informed predictions to navigate its environment optimally.[19] As such, the brain is not a passive receiver but an active generator of mental models and hypotheses, which it tests against incoming sensory evidence. This probabilistic model of perception traces its roots to the before-mentioned Hermann von Helmholtz's concept of "unconscious inference," now formalized within the framework of

[19] Ibid.: 131.

hierarchical Bayesian inference.[20] This model seeks to explain how sensations give rise to beliefs and guide actions. Key concepts within this approach, such as predictive coding and predictive processing, describe how the brain anticipates and interprets sensory data by generating predictions based on prior experience and adjusting these predictions as new evidence emerges.

According to Friston and Frith, "predictive coding" refers to a process in which higher cortical regions generate predictions about the representations in lower regions, effectively creating a top-down flow of expectations. These predictions are then compared against sensory input at lower levels, generating a "prediction error" that reflects discrepancies between anticipated and actual sensory experiences.[21] This error signal, representing the mismatch between expectation and reality, is subsequently sent back up to adjust and refine the higher-level representation, thus continuously updating one's cognitive framework in light of new experiences.

Friston and Frith further argue that neuronal activity encodes beliefs, or, more specifically, "probability distributions over states in the world that cause sensations."[22] This suggests that the brain actively infers causes behind sensory input, implying an external reality that generates these sensations. Once a sensation is perceived—such as a visual stimulus—it undergoes top-down processing and feedback integration, transforming sensory information into a conscious perception that is invariably shaped by prior expectations. This cycle of perception and expectation reflects a fundamental mechanism by which the brain operates: perceiving and acting are forms of inference, grounded in probabilistic predictions and expectations built from a lifetime of accumulated experiences.

Predictive coding and predictive processing are computational frameworks for understanding perception, in which the brain functions somewhat like a computer, applying background knowledge and expectations to interpret sensory data. Although this approach to interpretation is mechanistic, it still involves elements of expectation and context-based inference, albeit in a constrained sense. As originally proposed in Helmholtz's *Treatise on Physiological Optics* (1867), this theory suggests that perception is inherently inferential, relying on prior experiences to form interpretations of sensory input. The core of this process is modeled

[20] Ibid.
[21] Ibid.
[22] Ibid.

by the principle of Prediction Error Minimization (PEM), which assumes that the brain's access to the external world is indirect, mediated by layers of inference and prediction correction.

The PEM framework operates under the assumption that, because the brain is encased within the skull and insulated from direct interaction with the external world,[23] it relies on neuronal processes to generate interpretations. These interpretations draw on "priors," or probabilistic hypotheses formed from past experiences. When sensory input does not align with these predictions, the brain revises its model to reduce prediction error.

The brain's restricted access to the external world is mathematically represented by the Markov blanket, a boundary that delineates the limits of cognition. Andy Clark's interpretation broadens this boundary by incorporating aspects of the body and environment, suggesting that cognition involves developing low-cost, adaptive strategies for navigating the world. These strategies draw continuously on environmental structure and feedback,[24] which the brain integrates through action and sensory engagement.

Ultimately, this model implies that perceptions are not direct encounters with the world but are instead indirect reconstructions—corrections of predictions informed by accumulated experiences and prior representations.

Friston's PEM or "neuro-hermeneutics" has very little in common with hermeneutics. It is a model who sees the human brain as the seat of computational corrections of predictions. Friston model understands medical image interpretation as a Bayesian updating process, where the brain continuously minimizes prediction errors to refine its internal computation. However, this approach is overly reductive.

By reducing interpretation to a computational task of uncertainty minimization, Friston's model constrains the act of understanding to a process of probabilistic adjustment. In contrast, hermeneutics emphasizes that understanding is a culturally embedded exchange of opinion and understanding. In medical imaging, this dialogue includes discussions with colleagues and an engagement with institutional practices and knowledge.

Neuro-hermeneutics assumes that interpretation is a purely cognitive process, but hermeneutic philosophy insists that understanding is fundamentally embodied. Meaning is not passively extracted from an image but

[23] Hohwy (2016:259).
[24] Clark (2016: 191).

actively shaped through engagement with it. Radiologists do not work in isolation; their expertise is cultivated through years of training, professional intuition, and collaborative cooperation. Interpretation in radiology is rarely a solitary process—it involves second opinions and interdisciplinary discussions—hence, radiological interpretation is an ongoing refinement of understanding.

Rather than conforming to a rigid model of inference, the hermeneutic alternative embraces the fusion of horizons—an ongoing, open-ended, and contextual process where understanding is shaped through practical engagement. Interpretation is not a finite calculation but an evolving engaged interaction, reinforcing the fundamentally human and social nature of medical reasoning.

## 11.5 Extended Cognition

Embodied enactive cognition offers a valuable framework for analyzing cognitive processes in radiology, particularly in the context of radiologists' interactions with advanced imaging technologies. This discussion will begin by examining the concept of embodied enactive cognition alongside two complementary theories—postphenomenology and phenomenological hermeneutics—to assess how well these frameworks accommodate and elucidate radiologists' interactions with PACS and image display technologies. Specifically, this analysis will explore how compatible and well-suited these theories are for investigating the cognitive and interpretative demands placed on radiologists as they interact with X-ray, MRI, CT, and other imaging modalities.

A central question addressed here concerns the extent to which radiologists are both cognitively and bodily equipped to access and interpret the information revealed by these imaging technologies. As advancements in medical imaging continue, such as higher resolution images with enhanced pixel density, radiologists are increasingly challenged to engage with more detailed and complex visual data. In addition, developments in artificial intelligence (AI), particularly machine learning and deep learning (often branded as "deep medicine"), are being integrated into radiological practice with customized, locally trained algorithms to assist in image interpretation.[25] These AI-driven tools are gaining empirical validation, showing

[25] "Locally" means that the data used to train the algorithm is relevant for the population at hand.

that they can perform interpretative tasks at a level comparable to expert radiologists. Such innovations address perceptual and interpretive challenges inherent in radiological practice, yet they also introduce complexities. For instance, while increased pixel density enhances image quality and reveals details previously undetectable, it also generates visual noise, potentially complicating pattern recognition and interpretive accuracy. Radiologists must actively manipulate these image-reading tools to mitigate noise and optimize the visibility of critical diagnostic features.

The various cognitive theories that inform this discussion—embodied cognition, enactive cognition, extended cognition, situated cognition, and distributed cognition[26]—offer diverse perspectives on the mind-body relationship and the dynamic interplay between cognition and the environment. Embodied enactive cognition, in particular, provides a lens for understanding how the "body, brain, mind, and environment" form an integrated system in which cognition is shaped through interactions with the external world. This perspective posits cognition as an interactive process grounded in bodily and environmental engagement, in contrast to traditional cognitive theories that emphasize symbolic processing. Embodied enactive cognition emerged as a response to the limitations of symbol-based approaches in addressing real-world perceptual and interpretative problems.[27] It rejects a symbol-system model of cognition and instead highlights how human perception is inherently interpretative, continuously shaped by interactions with technological environments.

This theoretical framework challenges the efficacy of traditional cognitivist models—whether psychophysical or otherwise—to address perceptual issues specific to radiology or to offer cognitive solutions to such problems. Embodied enactive cognition thus underscores the interpretative nature of human perception, aligning well with the complex, interactive demands radiologists face when working with sophisticated imaging technologies. Consequently, this approach provides a robust foundation for evaluating radiologists' interpretive interactions with technology and offers insights into the cognitive adjustments needed as AI and imaging capabilities continue to evolve.

Embodied enactive cognition is inherently directed, contextual, and initiates human action. In fact, within this framework, cognition itself is activity. Human cognitive processes are not passive or isolated but are

[26] Shapiro (2014).
[27] Gill (2018).

actively oriented toward the environment, integrating perception and action into a continuous flow of engagement. As neuroscientist Laila Craighero emphasizes, "cognitive processes of all kinds are rooted in perception and action."[28] This form of cognition involves a directed interpretative or simulative activity, a process of meaning-making that emerges pre-consciously and is intimately tied to our bodily and historical embeddedness.

In her work, Craighero[29] highlights how recent advances in neuroscience reveal the motor cortex's central role in complex cognitive functions. The motor cortex, or agranular frontal cortex, comprises at least seven distinct areas, each with specific functions that contribute to motor control. Five of these areas, which receive cortical input from the parietal lobe, connect directly with the spinal cord and interact with the primary motor cortex. These motor areas are essential for analyzing diverse sensory inputs[30] (such as auditory, olfactory, and visual information processed along the optic nerve) and transforming them into coherent perceptions. Importantly, Craighero underscores that spatial perception is not managed by a single, multi-purpose region. Instead, the brain constructs multiple spatial scenarios, each related to particular classes of actions.

Of particular relevance to this discussion is Craighero's examination of "mirror neurons" and their function within the parietofrontal circuits.[31] These neurons provide a compelling explanation for the enactive link between perception and action. Mirror neurons activate not only when an individual performs an action but also when observing someone else performing the same action, thereby "mirroring" the other's behavior. Neuroscientists and cognitive scientists see these neurons as crucial for understanding others' actions and for learning skills through imitation—an insight highly relevant to radiological training, where practitioners acquire complex skills by observing and replicating expert techniques.

Craighero's findings align closely with the concept of "filling-in," where pre-conscious enactive processes allow individuals to respond effectively even with fragmented sensory input and memory. Notably, Craighero's research demonstrates that mirror neurons can activate even without a fully complete sensory input; some neurons fire in response to

[28] Craighero (2014).
[29] Ibid., 51.
[30] Ibid., 54.
[31] Ibid.

actions that are only partially visible, suggesting that the motor system has an inherent capacity for prediction and anticipatory action.[32] This aspect of mirror neuron function underscores the role of the motor cortex in visuomotor transformation, spatial coding, motor learning, and the anticipation of others' actions.[33]

These neuroscientific insights reinforce the enactivist view of human beings as self-individuating systems, a concept explored by theorists such as Di Paolo and Thompson. According to this perspective, self-individuation entails an organism's ability to generate and sustain itself through constant structural and functional adaptation.[34] This ongoing self-organization allows the organism to distinguish itself as an identifiable entity in relation to its surroundings. Such a view emphasizes that cognition is not merely a passive reception of stimuli but an active, dynamic process through which the individual continuously redefines and maintains their distinctiveness in interaction with the world.[35] This enactivist perspective thus provides a robust foundation for understanding how radiologists, as embodied cognizers, engage with complex imaging technologies, actively interpreting and interacting with the environment in ways that are deeply rooted in both perception and action.

Another compelling perspective on embodiment comes from Andy Clark's "extended mind" theory, which explores how humans use technology to expand cognitive capacities and reduce cognitive load.[36] Clark illustrates how individuals rely on tools and technologies—ranging from simple instruments like pen and paper to sophisticated systems like PACS and RIS, often integrated with MRI—to extend their cognitive reach, such as memory and perception, into the external material world. This extension of cognitive function highlights how technological integration can transform the scope and depth of human cognition.

Cathrine Hasse[37] further elaborates on these ideas through the lens of American philosopher Don Ihde's postphenomenological approach, which incorporates "extended," "embodied," and "hermeneutic" perspectives. According to Ihde, artifacts not only extend human perception and cognition but also mediate and reshape our interaction with the

[32] Ibid.

[33] Ibid.: 56.

[34] Di Paolo and Thompson (2014: 69).

[35] Ibid.: 68.

[36] Clark and Chalmers (1998); Magnusson (2009); Wheeler (2014); Hasse (2020).

[37] Hasse (2020: 254).

world. For example, adjusting to technology, such as wearing glasses, requires a gradual process of bodily and perceptual integration. Over time, as the user becomes accustomed to the glasses, they no longer feel intrusive; they become an extension of the body, providing seamless access to visual information. This form of technology use exemplifies what Ihde calls a "transparent" embodiment, where the tool becomes a naturalized part of perception, fading from conscious awareness as the brain adapts to its presence. By contrast, technologies that demand active interpretation—such as certain imaging tools—remain persistently "present" and resist such seamless integration. Here, cognitive extension requires a deliberate interpretive process, as the tool serves as an intermediary between the user and the world.[38]

Clark and Chalmers famously raised the question, "Where does the mind end, and the external world begin?"[39] In analyzing radiologists' interactions with PACS, Clark's theory of extended cognition shows us how instruments, tools, and even natural objects can function as extensions of human cognitive processes. Technologies such as machine learning and deep learning algorithms may actually serve as powerful extensions of radiological cognition. These systems, which utilize analogic and artificial neural network algorithms, are capable of self-correcting through extensive data processing. Though these algorithms do not replicate human experience or even the function of individual neurons, they provide computational support that extends human interpretative and perceptual capacities.

Embodied cognition in radiology involves interpreting the moment when technology-mediated patterns become recognizable. This process entails not only pre-conscious bodily and neural processing but also the cognitive extensions that tools provide. For example, when a radiologist uses PACS and RIS, the act of navigating images with a computer mouse exemplifies how tool use extends cognitive function. Emerging research suggests that the space immediately surrounding the body, known as peripersonal space,[40] is associated with a specialized system in the brain. Extensive experience with the computer mouse can even extend the auditory aspect of this peripersonal space, as visuomotor interaction through

[38] Ibid.

[39] Clark and Chalmers (1998: 1).

[40] Bassolino et al. (2010: 807).

the mouse also expands audio-tactile integrative space around the hand, indicating that multisensory processing dynamically adjusts to tool use.[41]

Cognitive extension involves more than merely influencing cognition through play or technology—it requires alignment among body, mind, and technology. This alignment optimizes the extended mind's functionality, as each element interacts to support complex cognitive tasks. In radiology, for instance, the hand that maneuvers the mouse shapes the task directed by the computer's deep learning algorithms, embodying a fusion of Hubert Dreyfus's concept of skill acquisition and embodied enactive cognition as defined by Shaun Gallagher.[42]

Gallagher[43] argues that cognition is distributed across the brain, body, and environment, and, along with Bassolino, he emphasizes that bodily processes contribute significantly to cognitive formation. Biological elements, such as organismic and emotional regulation, shape cognition durably, while, as Alva Noë stresses, sensorimotor coupling between the organism and environment further influences cognitive processes. These insights suggest that cognition is not confined to the brain alone but arises from the intricate interactions between bodily functions, perceptual engagement, and environmental structures. This distributed approach to cognition underscores how technologies used in radiology become integrated within this network of mind, body, and world, thus actively shaping interpretation in clinical practice.

Following Gallagher,[44] we adopt an inclusive interpretation of the extended mind that encompasses social and cultural dimensions.[45] Cognition does not solely occur within the brain; rather, it is supported by external resources that, as Gallagher argues, should be reliably available, frequently utilized, automatically endorsed, trustworthy, and readily accessible.[46] For instance, when one moves a computer mouse, it is not the brain itself that is extended but the cognitive process involving the movement. The motion of the hand activates signals in the mouse, which then interface with the computer—forming a cognitive loop between bodily musculature, the mouse, and the digital system. This interaction, Gallagher

[41] Ibid.: 810.
[42] Gallagher (2011a).
[43] Ibid.: (2011a: 66).
[44] Gallagher (2011b).
[45] Noë (2004).
[46] Gallagher (2011b).

contends, exemplifies how cognition can extend beyond the brain through tools and actions rather than merely "extending" the brain itself.[47]

Such cognitive extensions transform neural and mental processes as they are externalized and engaged. Research by Bassolino et al.[48] and Gallagher supports this view, emphasizing that extensions function alongside embodied, embedded, and enactive interpretation. Thus, the mind is enactively generated through interactions within both the social and physical environments.[49] Essential to this extended cognition is its interpretative nature—interpretation arises as an active process through which individuals create and understand meaning.

Interpretation, therefore, is neither simply descriptive nor procedural but event-driven. In the case of radiological image analysis, interpretative engagement involves myriad neural processes and often an intuitive recognition, as when a pattern becomes clear in consciousness. For example, a radiologist may recognize a lung nodule when viewing a PACS system image, a recognition that emerges as she engages the image while drawing upon her skill memory. The nodule is not recognized solely through rational analysis but also through perceptual attunement to the nuances of the grayscale image, a skill honed over years of experience. Her cognitive response is thus embedded in a deep sensory attunement to the technology and the image itself.

Similarly, a cardiologist inserting a catheter to clear an artery engages tactile, visual, and auditory senses to navigate to the heart, where he will remove plaque. This procedure demands careful sensitivity to both the patient's anatomy and the imaging technology that aids navigation. Entering a beating heart poses challenges that require the cardiologist to gauge the fit of the catheter to the artery size and to feel his way delicately to the target chamber. The work rests on systematic practice but draws heavily on skill memory. The cardiologist's cognition adjusts dynamically to obstacles such as arterial curves. His fingers, arm, tactile awareness, and visual monitoring coordinate with his internalized expertise, allowing him to navigate the catheter precisely. His bodily attunement and sensory responses occur largely outside conscious awareness, embodying a form of cognition that is spontaneous and intuitive, rooted in prior skill rather than deliberate thought.

[47] Ibid.: 8.

[48] Bassolino et al. (2010).

[49] Gallagher (2011b).

This embodied approach to cognition posits that the body, in conjunction with the brain, significantly shapes cognitive processes. Varela et al. define "embodiment" as the way cognitive processes are rooted in bodily interactions with the world, indicating that cognition is inseparably linked to physical engagement and enacted through bodily and environmental interaction. These interactions highlight the body's role in not only supporting cognitive tasks but in transforming them, underscoring that human cognition is fundamentally shaped by embodied and extended relationships with both social and physical contexts.

> By using the term *embodied* we mean to highlight two points: first that cognition depends upon the kinds of experience that come from having a body with various sensorimotor capacities, and second, that these individual sensorimotor capacities are themselves embedded in a more encompassing biological, psychological and cultural context.[50]

Embodied cognition explores the ways in which cognitive processes are influenced by and integrated with bodily interactions. While cognition includes conceptualizations and judgments, the body contributes to cognition through its motor system, sensory perceptions, and direct interactions with the environment. This situated nature of cognition suggests that individuals are both physically and cognitively embedded in their surroundings, which they actively engage with—visually, audibly, and through other sensory channels. Consequently, cognition is inherently situated within and shaped by the physical, social, and cultural contexts in which it occurs.

In the enactivist view, visual perception is not merely a passive intake of data but an interpretative and active exploration of the world, where understanding is expressed as a dynamic hermeneutic interaction between the body and environment. This perspective highlights that perception involves a preliminary, interpretative engagement that shapes what becomes conscious visual information. Interpretation, therefore, is a process of actively constructing and interacting with the world, embedded in bodily experience and grounded in the continuous interplay between perceptions, movement, and understanding.

[50] Varela, Thompson, Rosch (1993: 172–173).

CHAPTER 12

# Advancing the Theory of Understanding

Are our perceptions and interpretations merely indirect representations—constructed and adjusted against prior experiences via predictive processing—where the brain as a "prediction machine" has minimal direct access to the external world? Or do sensations provide a direct, construction-free correspondence with that world? There is reason to consider a middle path between these two extremes: a perspective that embraces the dynamic interplay of individual perceptual engagement with the environment, navigating a complex array of known and unknown elements, both directly and indirectly, while engaging in ongoing processes of sense-making and skill acquisition.

The prevailing debate between these stances often presents an absolutist dichotomy, with advocates on each side reluctant to concede any validity to the other. For strict constructivists, perception is a mediated, interpretative process; there is no immediate, relational meaning-making in real time. In contrast, proponents of direct perception deny that prior experiences influence perception as a corrective measure or contribute tentative meanings to sensory information.

In Bayesian predictive processing, perception is understood through the lens of indirect realism—an inferential process in which the brain continually updates and refines internal models to minimize discrepancies between perception and reality, based on computations of past experiences. Andy Clark, a prominent proponent of this view, argues that our



J. K. B. O. Friis, *Hermeneutics at the Intersection of Medical Technology*, https://doi.org/10.1007/978-3-031-88755-0_12

knowledge of the world is necessarily indirect, mediated by internal models that structure perceptions of the external environment hidden behind the "veil" of sensory data.[1] This perspective asserts that perception involves rapid cue-processing, where sensory bottom-up signals are continuously aligned with top-down predictions and lateral inferences about what is most probable based on prior knowledge and context.[2] This recursive interaction between bottom-up and top-down processes generates a perception that is neither purely direct nor fully constructed, but a dynamic, interpretive process shaped by both immediate sensations and past experiences.

The direct realist view that supports a dynamic and relational understanding of perception is exemplified by the sensorimotor theory within enactivism. This perspective views human perception as a self-organizing, autonomous process, where perception arises as an enactment of meaningful interactions between an organism and its environment. Rather than a passive process of merely acquiring sensory "data," this account, in my opinion, sees perception as actively involving an organism's embodied engagement with its surroundings, where the perceptual experience is not necessarily isolated from prior experiences, beliefs, social contexts, or emotional responses. A traditional account of direct perception assumes a straightforward acquisition of sensory cues, which would imply no "construction" or influence from these contextual factors. However, acknowledging the influence of prior experience and belief, even at a subconscious level, would seem to undermine the idea of a purely direct mind-world connection.

The notion of "construction" or a "background of perceptual influences" is thus complex. Degenaar and O'Regan, in their discussion of sensorimotor theory and enactivism, explore what constitutes perceptual consciousness.[3] They argue that perceptual consciousness is a form of awareness rooted in sensorimotor skills—directly engaged with the environment through sensory experiences across modalities, such as sight, sound, and touch, but excluding cognitive processes like thought, emotion, or internal psychological states. This exclusion is a critical point for preserving the theory as one of direct perception, as the presence of

[1] Clark (2016).
[2] Ibid.
[3] Degenaar and O'Regan (2015: 1).

construction-like processes or "representations" could theoretically disrupt the theory's commitment to directness.

However, if we set aside terms like "representation" or "construction" and examine what has been omitted, it becomes clear that beliefs, emotions, and pre-existing frameworks invariably influence perceptual engagement. These affective and cognitive elements act as intermediaries in all interactions between individuals and their environments, shaping perceptions in ways that may not strictly align with the idea of direct perception. Thus, even in enactivist frameworks, emotions, beliefs, and embodied experiences contribute to the interpretative process, suggesting that they are intrinsic to perceptual understanding and cannot be entirely separated from it.

In his analysis of belief systems, Avijit Lahiri[4] posits that beliefs are not isolated entities but instead form an intricate network of relational dynamics within an individual's mind. These beliefs are interdependent, with some reinforcing one another and others existing in a state of conflict. This structure gives rise to a belief system characterized by inherent tensions and dynamic interactions among its various components. Additionally, many beliefs reside within the unconscious mind, where they engage in complex interactions that frequently bypass conventional norms of rationality. Lahiri suggests that these unconscious beliefs—often absorbed passively and unexamined—can be inferred from behavioral patterns even though they remain unacknowledged. Operating involuntarily, they resemble hidden currents beneath the surface, influencing conscious thought and action in subtle yet significant ways.

From a functional standpoint, beliefs could be seen as a cognitive map that assists individuals in navigating the complexities and uncertainties of life. This map, as Lahiri describes, is not static; it is continuously revised through experience, trial, and error, and varies in reliability. Nevertheless, it serves as an essential navigational tool. We argue further that beliefs form a vital part of an individual's pre-understanding, playing a foundational role in processes of interpretation, as well as in both inductive and deductive reasoning. Lahiri also suggests, and I agree, that beliefs help shape individual psychological characteristics, creating unique approaches to engaging with the world. This dynamic interplay weaves together

[4] Lahiri (2020: 110–120).

beliefs, emotions, and moods, which are often rooted in physiological responses and largely operate beyond conscious control.[5]

In the interpretative process, beliefs—shaped by emotions and unconscious processes—serve as the bedrock for further cognitive activity. Pre-understanding, in this sense, may be likened to a "Bat Kol,"[6] a metaphor for a divine voice surfacing from within, guiding perception through the interplay of one's internal beliefs and external experiences. This interwoven tapestry of memories, lived experiences, and internalized knowledge enables an individual to interpret the external world through the lens of the self, with no clear separation between the two. What emerges naturally is a profound reality that reveals itself to the individual. This reality can be unraveled and harnessed to interpret the observed world, much like the creative processes of writers and composers, who lose themselves in moments of profound inspiration, guided by an inner voice that brings forth expression from the depths of the unconscious.

Sensorimotor theory positions understanding as a fundamental component of perception, asserting that perception is inherently bound to one's capacity to understand.[7] This is engagement occurring without an intermediary layer separating experience from the world; rather, experience itself is the act of engaging with and understanding the environment. Understanding thus becomes a form of "knowing-how," equipping—or "setting up"—an individual to act and interact effectively within their surroundings. It is a practical skill—one that involves discerning what to do and how to do it, and even the process of discovering new ways of interaction becomes a task rooted in understanding.

Central to sensorimotor theory is its focus on the perception of a three-dimensional world, emphasizing how vision operates within this framework. Through sensorimotor engagement, individuals possess an implicit understanding of how to interact with an object, whether near or distant, large or small. Beaton,[8] along with Degenaar and O'Regan,[9] attributes this to the structured nature of our perceptual abilities, which enable us to perceive and comprehend the significance of our surroundings. This approach underscores that perceiving is inherently practical rather than

[5] Ibid.

[6] Bat Kol is a Hebrew word and means the daughter of the Voice, bringing the word of God to us.

[7] Beaton (2016: 266).

[8] Beaton (2016: 267).

[9] Degenaar and O'Regan (2015: 1).

theoretical. Thus, perception does not entail a strict, objective representation of an object in a one-to-one correspondence but is instead shaped by the individual's perceptual capacities, which are structured to make sense of the world in a functional manner. These capacities, however, are not fixed; they are adaptable and can be recalibrated in response to environmental changes. Through experiential learning, we can shift our perceptual abilities, effectively learning to "see" differently by engaging with new perceptual affordances.

A notable illustration of this adaptive perceptual skill is Goethe's discovery of the *os intermaxillare* (the premaxilla bone) in the human skull. For a long time, it was assumed that this bone did not exist in humans. However, through a focused, imaginative study of skull structures across primate species, Goethe was able to refine his perceptual skillset, discerning previously overlooked patterns in human skull structure. This discovery emerged not from passive observation but from a practical, engaged interaction with the skull as an object of study. Beaton emphasizes that "our perceptual skills are not just hard-wired, automatic, or impervious to change"; instead, as Noë elaborates, such perceptual adaptation requires active attention, purpose, and interaction.[10]

This process of adapting one's perceptual skills—termed "re-skilling" by Beaton—involves the application of fundamental perceptual knowledge, which he describes as a "species of knowledge" intricately connected to explicit, structured understanding.[11] This concept of re-skilling is particularly relevant to fields such as radiology, where practitioners must refine their perceptual abilities to interpret complex, technologically mediated 2D and 3D images. For radiologists, this entails learning to identify new patterns and structures, requiring both an engagement with novel affordances and the continual development of embodied perceptual skills tailored to their domain.

The question of direct realism remains unresolved, particularly in the context of how perception interacts with cognitive structures. Beaton, drawing on Immanuel Kant's ideas, suggests that while we may see objects directly, our perception is inseparably bound to the meanings we attribute to them. For example, when we see a pipe, we do not perceive it as a purely objective form devoid of meaning; instead, we immediately recognize its functionality and purpose. Eleanor Rosch and her colleagues

[10] Noë (2004).
[11] Beaton (2016: 268).

describe this phenomenon as a process of basic-level categorization, grouping objects like pipes, chairs, and tables[12] into categories that are cognitively fundamental. Objects within these categories are recognized based on shared, perceivable characteristics, and individuals interact with these objects using similar motor actions. This level of categorization, therefore, represents the convergence of attributes that are most accessible to perception and action.

The organization of such categories is not predetermined;[13] rather, it arises from the sensorimotor structure of the human body and its interpretative capacities, as well as the range of interactions this structure supports. For example, in the context of radiology, medical images are categorized and interpreted according to image-schemas that radiologists develop through repeated exposure to certain types of patterns in medical imagery. This concept reflects the ideas of Varela,[14] who, building on Piaget's work and later developments by George Lakoff and Mark Johnson, proposes the existence of "embodied cognitive structures" that shape perception. In this framework, reality—as experienced by a radiologist—is not an objective construct but rather one that is dependent on the perceiver. The interpretation and perceived "reality" within a radiological context are shaped by the cognitive structures embodied by the radiologist, which means that the objectivity of interpretation is intrinsically tied to the perspective and expertise of the individual perceiver.

This perceiver-dependency, however, is not arbitrary or subject to personal whims. Rather, as Varela would argue, what constitutes "radiological reality" is fundamentally intertwined with the embodied cognitive framework of the radiologist. In the clinical context, perceptual interpretation is shaped and guided by this embodied structure, which is further modulated by environmental factors specific to the practice of radiology. Thus, perceptual interpretation is not merely a matter of passively observing visual features; it is an active, perceptually guided process where recurrent sensorimotor patterns, ingrained through experience, enable contextually appropriate interpretations of medical images.

The distinction between "perceptually guided action" and "visual feature extraction" is essential to understanding this interpretative process.[15]

[12] Rosch et al. (1976).
[13] Varela (1999: 16).
[14] Ibid.: 13.
[15] Ibid.: 12.

Perceptually guided action refers to an active engagement with the image content, shaped by the radiologist's learned patterns of interaction with similar images. This approach suggests that interpretation is inseparable from the radiologist's embodied cognitive structures, as these structures underpin and shape the reality of the interpretation. In other words, the reality perceived and interpreted by the radiologist is not merely derived from passive observation but emerges from a complex interplay between perceptual capabilities and sensorimotor experiences, which together define how the radiologist interacts with and interprets medical images.

In *Ethical Know-How*, Francisco Varela offers a compelling exploration of how perception is intricately linked with bodily movement and the interplay of multiple sensory modalities. He illustrates this concept through two key examples.

The first example draws on a study by Richard Held and Alan Hein,[16] who raised kittens in a dark environment, allowing them exposure to light only in controlled circumstances. The kittens were divided into two groups: the first group could move freely within the confines of a harness, while the second group remained entirely passive, seated in a basket attached to the same carriage. Although both groups had identical visual exposure, only the active kittens developed normal perceptual abilities. When later released, the passive kittens exhibited severe perceptual impairments, behaving as though blind—they stumbled over objects and struggled to navigate edges. Varela interprets these findings as evidence that perception is not a matter of passive visual pattern extraction; instead, it emerges through the interaction of visual stimuli with self-directed bodily movement.

The second example, a study by Paul Bach-y-Rita,[17] involved a device designed to aid blind individuals. The device used a video camera to transmit images as electrical vibrations to the skin. Crucially, the tactile feedback only produced coherent images when the person moved; without movement, no perceptual image was formed. This also suggests that perception depends on the dynamic, embodied movement of the perceiver.[18]

These examples underscore an interpretive process in perception that cannot be fully explained as direct sensory input alone. Rather, they reveal that perception is actively constructed through bodily engagement with

[16] Varela (1999: 14).
[17] Ibid.
[18] Ibid.

the world. The formation of "images" in the mind relies on movements that integrate multiple sensory channels, which gradually transform and re-construct sensory data into a coherent subjective reality. Thus, perception emerges as an adaptive interaction between the individuals and their environment—a synthesis of embodied experience rather than a passive reception of visual data.

The process of recontextualizing memory and experience can be understood as an active, reconstructive, and self-modifying endeavor in the ongoing formation of meaning. Human beings do not encounter the world with a "blank slate" as they navigate daily life. Instead, our perception is grounded more in the viability of experiential interpretation than in precise representational accuracy. As we engage in routine activities, we implicitly and contextually recycle and adapt meanings. Although creativity plays a role in constructing and sustaining meaning, this meaning is not arbitrary. Rather, it is shaped by circular feedback mechanisms inherent in both neural and cognitive interpretive processes. Our engagement with reality involves a dynamic, cognitive process that blends new stimuli with prior knowledge and experience, described by Maturana and Varela as "structural coupling." This term refers to the interdependent relationship between individuals and their environment, in which cognition is both shaped by and responsive to contextual constraints and possibilities.[19]

Structural coupling also involves an openness to shared perspectives—an integration of others into our interpretive frameworks. Human existence is inherently relational, with every action emerging from a set of preceding conditions that afford specific possibilities for perception and response. In this view, nothing exists in isolation; our environment continuously shapes and is shaped by our actions and interpretations. For example, identifying patterns within mediated contexts, such as X-ray analysis, is an act of meaning-making grounded in understanding. This interpretive process involves a recursive and transformative circularity, where each interpretative cycle's outcome becomes the input for the next, allowing for progressively refined interpretations. Through this dynamic cycle, continuity is maintained across experiences, intertwining past and present in each moment of comprehension.

One of the primary challenges for theories of direct perception lies in addressing the role of memory, experience, and feedback mechanisms—particularly the role of consciousness in creating meaning within the world

[19] Maturana and Varela (1980, 1987).

we inhabit and interact with. As Pickering[20] suggests, humans engage with their environment through creativity and metaphor, as adaptive, "open strategy" organisms. This means that while innate abilities are essential, they are insufficient for adapting fully to specific environmental contexts. Human evolution requires not only biologically inherited capabilities but also the development of new skills to navigate complex environments. Humans' innate and acquired perceptual capacities include metaphorical perception and mimetic action, which extend beyond mere biological predispositions and are essential to our flexible and adaptive interaction with the world.[21]

Pickering addresses a compelling dimension of human interpretive capacity, particularly through his discussion of mimesis, or the ability to act "as if" something were other than it actually is. This capacity for "as if" reasoning underlies a uniquely human form of communication that is not present in animals. According to Pickering, humans generate a wide array of vocalizations, gestures, and bodily movements—primarily through language—that facilitate shared attention to objects and situations. This ability to regard an object or context symbolically, as representing something other than its immediate reality, enables humans to find novel, often metaphorical uses for these objects. Unlike animals, who operate within the constraints of an instinctive behavioral repertoire, humans can detach objects from their standard meanings and reimagine their significance. This capacity to examine, reinterpret, and modify behaviors, artifacts, and symbolic representations, such as those found in art and images, exemplifies the distinctiveness of human perception and interpretative interaction.

Human beings partly lack direct perceptual access to the objective reality of things and events. Instead, the evolution of modern humans has been a prolonged process of cultural adaptation, which has shaped and refined our cognitive and practical skills in tandem with the development of technologies and tools. These cultural artifacts and technologies expand human interaction with the world and mediate our perception, transforming raw sensory input into complex interpretive acts. In many respects, the world we navigate is constructed by human cultural practices. Activities such as molecular biomedicine, radiology, astrophysics, and even driving involve highly developed symbolic systems and embodied cognitive practices, resulting in a distinctive, mediated perception. This perception is not

[20] John Pickering (2016: 280).

[21] Ibid.

a straightforward encounter with reality; it is an interpretive act—a process of seeing the world "as if" rather than "as is," as Pickering notes.[22]

Engaging with objects and tools, like an X-ray image or a car key, thus entails more than mere sensory experience. It involves culturally conditioned interpretation, which is grounded in continuous cycles of interpretation and adaptation. This interpretive dimension of perception adds a metaphorical layer to human experience, shaped by historical, cultural, and practical contexts. Understanding how to interact with these objects is not simply a matter of receiving sensory data but requires culturally informed intentions and a tacit comprehension of symbolic meaning.

## 12.1 Skills in the Making

Hubert and Stuart Dreyfus's exploration of skillful coping and the foundational background of knowledge offer a nuanced framework for analyzing how radiologists engage with medical images and imaging technologies.[23] Dreyfus differentiates between two primary forms of intelligent behavior: the process of learning and the execution of skillful action. Building on Merleau-Ponty's *Phenomenology of Perception*, Dreyfus employs two pivotal concepts: the intentional arc and maximal grip. The intentional arc describes the dynamic, evolving relationship between a radiologist and their experiential background, reflecting how expertise is internalized as a set of dispositions to act within a specific domain.[24] As radiologists advance in their practice, they continuously recontextualize and reorganize previous experiences, honing their skills to respond adeptly to new tasks or diagnostic challenges. This skill acquisition process is procedural and involves ongoing sensorimotor activity, enabling radiologists to discern specific patterns—such as the subtle presence of a cancerous nodule on an X-ray—that may go unnoticed by an untrained observer. Through this refined perception, the nodule emerges as a gestalt, an organized whole that appears spontaneously to the expert eye.

The memory associated with skillful coping is distributed across multiple neural maps, involving subcortical regions such as the hippocampus, basal ganglia, and cerebellum. Signals from the external environment are integrated into this broad mapping process, producing responses that

[22] Ibid.

[23] Dreyfus and Dreyfus, with Tom Athanasiou (1986: 15–36).

[24] Merleau-Ponty (1962).

guide bodily movement. Edelman and Tononi propose that this global mapping[25] is a dynamic, temporally sensitive structure that adapts according to behavioral demands. In radiological image analysis, reentrant circuits among local neural maps coordinate sensory inputs with motor responses, making gestalts visually perceptible. Recognition of recurring patterns, such as identifying a cancerous nodule, is facilitated by the reactivation of degenerate neural pathways, where similar perceptual outputs are produced from varying neural configurations. This mapping and selection of preferred pathways occurs largely at a tacit level, preceding conscious recognition of the gestalt within the X-ray. Thus, the radiologist's skillful interaction with medical images exemplifies a deep, embodied expertise that is both automatic and experientially grounded.

The concept of maximal grip, which Dreyfus adopts from Merleau-Ponty, describes the body's intrinsic drive to adjust its actions in ways that bring a given situation closer to an optimal gestalt. For radiologists, achieving this optimal gestalt involves the fine-tuned engagement of localized neural mappings and preferred pathways that evolve through continuous interaction with imaging technologies. Notably, neither intentional arc nor maximal grip relies on explicit mental representations. Instead, the process of learning a skill, such as interpreting radiological images, initially involves a series of repetitive actions. With time and practice, these actions become embodied, transforming into automatic, second-nature abilities that are seamlessly integrated within the radiologist's perceptual and cognitive framework.

In other words, Maximal grip refers to the body's inclination to adjust its position and actions toward an optimal view—an adjustment that brings a given image or detail into sharper focus. For radiologists, achieving maximal grip means synchronizing their actions and perceptions with the imaging technology they use, responding fluidly to each new set of image data.

How, then, does one become an expert radiologist or skilled image interpreter? According to Dreyfus, Merleau-Ponty, and Edelman & Tononi, skill acquisition is a transformative process shaped by interactions with specific objects (like radiological images) and contexts (such as medical training environments). These repeated interactions redefine how objects and situations reveal themselves to us. Edelman & Tononi suggest that memory itself is in constant reorganization, and as one develops

[25] Edelman and Tononi (2000: 96).

expertise, the relationship between the individual and their world undergoes a corresponding transformation. Dreyfus outlines five distinct stages in the progression from novice to expert radiologist, emphasizing that becoming an expert interpreter requires not only repeated exposure to diagnostic images, like lung nodules, but also proficiency in navigating advanced technologies, including PACS and electronic patient records. A key skill in this development is the ability to filter out technical noise in images, which is crucial for enhancing image quality and facilitating the recognition of optimal gestalts.

In the initial, novice stage, learners focus on context-free, decomposable features within the task environment—features that they can identify even without domain-specific experience. At this stage, learning is heavily representational and rule-based; the novice depends on clearly defined attributes and methodological guidelines. These methods gradually become embodied through repeated physical actions, with tasks like coordinated hand and finger movements evolving into automatic, sensorimotor routines. As the learner progresses to Dreyfus's second stage, they enter a training phase in which the expertise of their instructors plays a crucial role. Here, students are "found" within the guidance of seasoned practitioners whose embodied interpretive habits direct them. In other words, the role of mentorship becomes crucial; experienced radiologists model diagnostic techniques, imparting interpretive habits that novices mimic and internalize. This modeling is instrumental in helping trainees develop an interpretive "feel" for diagnostic images, which is further refined as they confront cases that are more complex.

The novice radiologist must also cultivate an individualized, systematic approach to image interpretation. This process requires reorganizing and adapting existing perceptual and cognitive habits to meet the evolving demands of radiological analysis. At this stage, the internalization of interpretative skills remains in progress; the necessary reorganization of memory and perceptual frameworks is ongoing, forming the foundation for the transition from rule-based learning to embodied expertise, where interpretative skill becomes second nature.

The next developmental phase in radiological expertise involves managing a significant influx of information. As the trainee is exposed to extensive new data, they must hone both their technological proficiency and interpretative abilities by integrating emerging mental models. This stage challenges learners to navigate the complex interactions between radiological images and the technologies that generate them. By the fourth stage, Dreyfus suggests that reliance on explicit rules and procedural principles diminishes, giving way to nuanced situational distinctions and instinctive responses. As expertise deepens, analytical thinking is gradually supplanted by a more intuitive, immediate approach to interpretation.

In the fifth and final stage—the two Dreyfus's expert phase—the radiologist achieves a level of mastery marked by highly refined perceptual discriminations. This capacity for fine distinctions sets the expert apart from the proficient performer, enabling rapid, intuitive responses that define genuine expertise. At this level, the radiologist's interpretative processes are not only highly practiced but also deeply integrated with technological tools, requiring advanced motor-sensory coordination to interact effectively with radiological imaging systems. One essential task at this stage is the elimination of extraneous noise from images, a process that greatly enhances diagnostic accuracy.

Modern radiological modalities, such as Magnetic Particle Imaging and Superb Microvascular Imaging, represent the forefront of medical technology. These sophisticated machines, comprising thousands of components with specialized functions, convert particle-level radiation data into images that can be visually perceived and interpreted. This transformation involves multiple layers of processing, through which raw radiation data is reconstructed into recognizable visual elements. Beyond these functional layers, physical layers—consisting of mechanical parts, electronic systems, and user interfaces—mediate the relationship between radiologist and patient, adding complexity to the interpretative process. Each hospital may customize its imaging modalities to address specific clinical requirements, further complicating standardization. To optimize these technologies, specialist radiographers and radiologists, known as "super-users," rigorously test and fine-tune the machines, adjusting features to minimize noise and enhance specific functionalities relevant to their subspecialties.

Since medical images are digital, noise is, as previously mentioned, an inherent challenge that degrades image quality, hindering the accurate visualization and analysis of clinical features.[26] Noise introduces artifacts that can result in false positives or false negatives, thereby impacting the reliability of diagnostic interpretations. Aspiring radiologists must learn to

[26] Goyal et al. (2018: 1227).

manage this issue by applying various denoising techniques, a critical skill in radiological practice. Noise can originate both during the image acquisition process and during post-processing. For example, computed tomography (CT) images are especially susceptible to noise due to their dependence on reconstructions from millions of discrete detector measurements.[27] Similarly, magnetic resonance imaging (MRI), while non-invasive and capable of generating highly detailed anatomical images, is prone to complex types of noise, such as Rician noise (from electronic interference), Gaussian noise (stemming from MRI hardware components like coils and circuits), and Rayleigh noise[28] (from thermal motion causing radiofrequency emissions). Each type of noise reduces image contrast, thereby obscuring both qualitative and quantitative details in the images. Mastery of denoising techniques is thus essential across imaging modalities, with each subspecialty requiring a specific skill set to maintain interpretative accuracy.

Much like how a hand instinctively adjusts to grasp an object, the radiologist adapts fluidly to each image, learning to operate levers, buttons, and settings of their technology. This practice is not simply a technical operation but an embodied integration of interpretive processes within a dynamic system of interactions, incorporating movement, gestures, and a deep engagement with the technological environment.

As outlined in Dreyfus & Dreyfus's stages of skill acquisition, expert radiologists exhibit a highly adaptable response to situations and objects within their practice. The entire environment—including the PACS, the radiological images, and the electronic patient records—demands a specific mode of interaction. Radiologists must not only attune themselves to the visual data but also to the technical system as an interconnected whole. There is no definitive rule dictating when, or if, interpretations are necessary; in such scenarios, the expert radiologist becomes fully immersed in the task. This immersion dissolves the distance between perceiver and perceived; the radiologist's focus is exclusively on the image of the injury or condition being analyzed. According to Dreyfus, the expert radiologist does not deliberate over actions but instead responds instinctively, continuously adapting to the unique demands of each case, each of which is subtly different from previous ones.[29]

[27] Barrett and Keat (2004: 1680).

[28] Goyal et al. (2018: 1234); Gudbjartsson and Patz (1995: 910–914).

[29] Dreyfus and Dreyfus, with Tom Athanasiou (1986: 15–36).

## 12.2 Interpretation and Memory

The processes of knowing, perceiving, and acting are fundamentally rooted in the continuous, dynamic interaction between the body, mind, and environment. Rather than operating in isolation, these processes function in an interdependent manner, with many of their activities unfolding prior to conscious awareness. This perspective challenges traditional views that emphasize rational deliberation as the primary mode of human cognition, offering instead a more comprehensive understanding of knowing as a participatory, embodied, and adaptive process.

Central to this argument is the notion of tacit knowledge, a concept championed by philosopher Michael Polanyi. Polanyi contends that all knowledge, no matter how explicit it may eventually become, emerges from a foundational substratum of tacit knowledge. This tacit dimension includes the unspoken, intuitive, and embodied understandings that underpin human engagement with the world. For example, a chess master may not be able to fully articulate the logic behind a particular move, yet their intuition is shaped by years of embodied practice and pattern recognition.

This view is further developed by Mark Johnson's conceptualization of knowing as an interactive, embodied process between the organism and its environment.[30] For Johnson, cognition is not a disembodied, abstract activity but a lived experience deeply enmeshed in the sensorimotor activities of the body. Knowledge is formed through continuous interaction and feedback between individuals and their surroundings. This dynamic process underscores the exploratory and experimental nature of knowing, where action and perception co-evolve in real time, often without the intervention of conscious deliberation. Johnson's emphasis on the embodied mind challenges Cartesian dualism, asserting that cognition is inherently situated within the lived experience of the body.

The role of habitual action provides further support for the interconnected nature of knowing, perceiving, and acting. Human beings develop ingrained habits, such as valuing, desiring, and acting, which guide their engagements with the world. These habits are not static; rather, they are subject to ongoing refinement as individuals encounter new experiences and challenges. Habits serve as adaptive mechanisms, enabling people to navigate their environments efficiently while still allowing for revision and

[30] Johnson (2017).

growth. The continuous process of evaluation and restructuring of experiences and values highlights the fluidity of human cognition and action. Through the repeated enactment of these habits, individuals cultivate practical wisdom, or *phronēsis*, which allows for the nuanced adjustment of actions in response to changing circumstances.

Neuroscientific perspectives on consciousness, particularly the work of Gerald Edelman and Giulio Tononi,[31] offer further evidence for the distributed and non-unified nature of human thought and action. Their theory of neural Darwinism proposes that consciousness arises from the dynamic selection and reconfiguration of neuronal groups within the brain. Rather than being a monolithic, unifying force, consciousness is a constantly shifting, distributed process. Multiple streams of sensory input, affective states, and cognitive processes operate in parallel, often outside the scope of conscious awareness. This decentralized perspective aligns with Polanyi's and Johnson's views on the tacit, embodied, and exploratory nature of knowing. The brain's capacity for parallel processing enables fluid adaptation to complex environments, reinforcing the notion that human thought and action are not solely the product of reflective, conscious deliberation.

Philosophers such as Mark Johnson, George Lakoff, and Francisco Varela offer an integrative perspective on the interplay between mind, body, and world, challenging traditional representational theories of cognition. They propose what may be termed a "hermeneutic framework" of pre-conscious coping and action, which is especially relevant in expert practices such as radiology. In this context, philosopher Yu Zhenhua,[32] drawing on Polanyi's insights, emphasizes that all knowledge is anchored in tacit, pre-conscious understanding. For instance, a radiologist's engagement with image analysis is informed by an extensive background of unexamined knowledge. This background knowledge enables automatic responses such as pattern recognition, edge detection, and object identification—cognitive activities that subtly inform the radiologist's visual engagement. Without this tacit background, the fluidity of the radiologist's analysis would be disrupted, rendering effective interpretation impossible.

Hubert Dreyfus's work on embodied expertise provides further support for this argument. Dreyfus illustrates how a radiologist's capacity to

[31] Edelman and Tononi (2000).
[32] Yu Zhenhua (2003).

act effectively without overt reflection on background knowledge underscores the necessity of such knowledge receding into the unconscious. This background is composed of acquired habits, non-conceptual coping skills, customs, and practical frameworks that shape perception and action. Dreyfus highlights the duality of knowledge, distinguishing between intellectual expertise (recognizing patterns within images) and practical proficiency (skillfully operating imaging technologies). This duality echoes Polanyi's notion of tacit knowledge, which manifests both intellectually and practically. Thus, a radiologist's expertise relies on the combined mastery of visual pattern recognition (connoisseurship) and technical skill (proficiency with PACS technology).[33] These implicit and context-dependent bodily, attitudinal, social, and cultural preconditions form a network of tacit knowledge essential for meaningful action.

Memory also plays a central role in this framework, functioning as an ongoing feedback loop between localized cognitive maps and global memory structures.[34] This continuous interaction between localized maps—such as those for sensorimotor, color, and retinal data—and global memory structures is fundamental to the construction of meaning. Neurologist Oliver Sacks highlights this dynamic nature of memory, noting, "When we open our eyes each morning, it is upon a world we have spent a lifetime learning to see. We are not given the world; we make our world through incessant experience, categorization, memory, reconnection."[35] This perspective reveals that memory is not a fixed repository of past events but a flexible, evolving network of relationships between perception and experience. Each new experience reshapes the memory structure, actively influencing the recollection and interpretation of past events. The result is a dynamic process of ongoing reconfiguration, with memory serving as both a product and a driver of adaptive knowing.

Memory, as argued, is an open, dynamic system characterized by ongoing recategorization and reconstruction.[36] While specific details within memory may remain imprecise, it is evident that neural maps do not offer direct representations of objective reality; rather, they approximate it through subjective, experience-driven reinterpretations. For radiologists, in particular, these memory maps form a crucial interpretive framework

[33] Ibid.
[34] Barry (1997); Johnson (2017).
[35] Barry (1997).
[36] Barry (1997); Edelman and Tononi (2000).

that informs and shapes their perception of complex images. As Johnson notes, these recurring adaptive patterns of interaction with the environment do not serve as internal representations but are essential to the ways in which humans—and other species—engage with the world.[37] In radiology, these embodied memory patterns allow experts to interpret images intuitively, relying on the implicit knowledge embedded in their cumulative experience and background.

A global memory map, therefore, is a complex, adaptable network composed of interconnected local maps, functioning as the cognitive substrate that supports unconscious and pre-conscious brain activity. When specific neural circuits are activated, they elicit particular responses that align with previously established patterns. This mechanism arises from the repeated coupling of mental or physical actions with specific neural pathways, reinforcing these associations over time.[38] Without this structured background of memory and accumulated experience, individuals would lack the foundational framework to navigate their lives[39]—there would be no interpretive "light," as Heidegger metaphorically describes. To fully appreciate hermeneutics as embodied interpretative cognition, it is essential to recognize the inherently creative nature of memory.

Neuroscientists Gerald Edelman and Giulio Tononi conceptualize memory as a dynamic, multi-layered system consisting of a vast repertoire of neural circuits. These circuits adapt by modifying synaptic connections in response to incoming signals and value-driven constraints, increasing the likelihood of reinforcing or recategorizing beneficial or adaptive responses.[40] Thus, memory should not be seen as a mere archival record but as an interpretative, adaptive process that sustains access to past experiences while reshaping their meanings. Francisco Varela highlights that cognitive processes continually draw upon memory, often beneath the level of conscious awareness. While some tasks engage our attention directly, others—such as the selective encoding of relevant memories or the formation of memory itself—occur beyond conscious perception. Memory thus operates as an active, self-modifying system that mediates our interaction with and interpretation of the world around us.[41]

[37] Johnson (2017).
[38] Edelman and Tononi (2000: 97).
[39] Ibid.
[40] Ibid.: 98.
[41] Varela (1999: 37).

In the domain of visual perception, the retrieval—or more precisely, the recontextualization—of memory plays a critical role in forming meaning-rich perceptual experiences. The brain exhibits distinct responses when engaging with three-dimensional environments as opposed to two-dimensional representations, with different neural regions activated to meet these varied perceptual demands. A 2011 study[42] highlighted this differentiation, revealing that two-dimensional images of objects triggered strong repetition effects within traditional object-selective cortical regions across both ventral and dorsal visual processing pathways. In contrast, when participants were presented with three-dimensional objects, these repetition effects were markedly diminished, if not entirely absent. These findings indicate that the neural processes underlying real, three-dimensional object perception may diverge considerably from those engaged during the interpretation of two-dimensional representations.

To deepen our understanding of this visual processing distinction, we consider Francisco Varela's model,[43] which conceptualizes neurons as an interwoven network of overlapping ensembles that generate coherent configurations based on the visual stimuli received. Upon observing an image, visual impulses move from the retina to the dorsal thalamus, and then from the thalamus to the visual cortex. A significant aspect of this process is the extensive feedback loop, with more neural connections flowing from the cortex back to the thalamus than in the initial feedforward direction. Varela posits that this feedback system is critical for integrating memory with contextual and value constraints, ensuring that perception and memory operate in a dynamic, interdependent manner—a mechanism particularly essential for a radiologist's visual interpretation.

Further research by DeYoe and Van Essen[44] reinforces the notion that visual perception is composed of a rich array of modalities, encompassing attributes such as shape, size, rigidity, surface properties (e.g., color and texture), spatial relationships, and movement within three-dimensional space. Varela further explains that these diverse elements of vision emerge from multiple, concurrent subnetworks within the brain, which work synchronously to produce a unified perceptual experience.[45] Visual signals are exchanged iteratively between distinct brain regions, coalescing into a

[42] Snow et al. (2011).
[43] Varela (1999: 45).
[44] De Yoe and Van-Essen (1988).
[45] Varela (1999: 48).

coherent gestalt—a perceptual pattern that becomes accessible to conscious awareness and imbued with meaning. This process exemplifies what may be termed "embodied unconscious processing": a form of interpretative neuro-processing that pre-consciously integrates visual inputs with memory, yielding a tentative outcome shaped by context and experience. Only selectively processed units of these pre-conscious interpretations reach conscious awareness, underscoring the intricate, interpretative nature of visual memory integration.

Varela, Thompson, and Rosch, drawing from thinkers like Heidegger and Gadamer, suggest that knowledge arises from continuous interpretation, deeply rooted in our biological embodiment and lived experience within a shared framework of consensual actions and cultural history.[46] In this perspective, Merleau-Ponty argues that humans exist in a dialectical relationship with their environment, simultaneously shaping and being shaped by it.[47] This reciprocal process fosters interpretative attunement, adaptation, and coping—key components of meaning as a lived, dynamic experience.

Gallagher[48] further posits that understanding perception and meaning-making requires both first-person phenomenological experiences and neuroscientific insights. He underscores the significance of mirror neurons, initially discovered in macaque monkeys and later identified in humans within areas like the premotor cortex and Broca's area.[49] Mirror neurons activate not only when an individual performs a motor action but also when observing the same action in others, thus bridging visual perception and first-person motor understanding.[50] This intermodal connectivity suggests a biological basis for empathy and shared understanding, highlighting how perception is informed by embodied, first-person knowledge.

The relationship between memory and propositional truth, however, is highly contested. Elizabeth F. Loftus, a pioneering researcher in memory studies, has demonstrated through extensive work that eyewitness testimony, long considered reliable, is often flawed. Her findings have led to significant reforms in legal practices, particularly in cases heavily reliant on eyewitness accounts. Following Loftus's groundbreaking work, numerous

[46] Varela et al. (1993: 149).

[47] Merleau-Ponty (1983: 13).

[48] Gallagher (2004: 3).

[49] Ibid.: 9. Mirror neurons awakened an intense interest among cognitive researchers during the first decade of the 2000s.

[50] Ibid.

studies have shown that memories can be highly malleable; in controlled experiments, participants have been led to "remember" events that never actually occurred. Through post-event misinformation, memory traces can be altered, causing individuals to falsely believe in the veracity of fabricated experiences. Laney and Loftus,[51] for instance, were able to implant false memories in 23.6% of 165 participants, convincing them they had experienced one of three entirely fictitious events. These implanted memories often included sensory and emotional details, underscoring the convincing nature of such fabricated recollections.

Loftus explains that memory undergoes a recoding process wherein new information merges with original memory traces, rendering the initial memory less accessible and often indistinguishable from modified details. She observes that "misleading information can turn a lie into memory's truth," enabling people to believe they have seen or experienced events that, in reality, never took place.[52] These artificially constructed memories can become as compelling as genuine recollections, blurring the line between original experiences and reconstructed narratives.

This understanding suggests that knowledge itself is a reconstructive process. Memory, similarly, operates through continuous reconstruction. Memory does not reproduce exact replicas of past experiences; rather, it reconstructs them in ways that appear accurate but differ from the original events. Often, these subtle changes go unnoticed by individuals, reinforcing the idea that knowledge and memory are fluid, evolving processes. Knowing is not a static state but a dynamic, unfolding activity.

Visual perception operates in a comparable manner, merging external stimuli with experiential knowledge to create meaning in a given moment. This phenomenon is observable in radiology, where intra-variability occurs: if a radiologist examines the same image at different times, the interpretations may vary due to shifts in perceptual and contextual factors. This variability highlights how perceptual and interpretive processes, informed by embodied knowledge and memory, interact dynamically to shape meaning, further illustrating the inherently reconstructive nature of human interpretation.

Taken together, these perspectives reveal a more holistic and dynamic understanding of knowing, perceiving, and acting. The body, mind, and environment operate as an integrated system, with tacit knowledge,

[51] Laney and Loftus (2008: 511, 513).

[52] Loftus (1992: 122–123).

embodied experience, and adaptive habits driving much of human activity. The experimental and exploratory nature of these processes allows for continuous learning and adaptation, even in the absence of conscious awareness. This perspective challenges the traditional emphasis on conscious, rational thought as the primary driver of human knowledge and action. Instead, it presents a more comprehensive model of cognition rooted in embodied, habitual, and distributed processes. Knowing, from this perspective, is not a static possession but an ongoing, participatory process of becoming attuned to the world. By synthesizing, we arrive at a rich, multifaceted view of human cognition as a dynamic interplay between the body, mind, and environment.

## 12.3 Summary and Conclusion

In concluding, we return to the intricate web of interaction among knowing, perceiving, and acting—processes fundamentally rooted in the seamless, dynamic relationship between body, mind, and environment. These processes often begin well before conscious awareness, underscoring that much of human cognition operates at an implicit, unspoken level. The concept of tacit knowledge proposes that all knowledge springs from an unarticulated base, while knowing is an embodied, continuous interaction between an organism and its surroundings. Together, these perspectives illustrate how knowing is more interpretively experimental and exploratory than fully conscious or deliberate.

This tacit dimension becomes central to expertise, particularly in radiology, where professionals rely on a vast, background knowledge that informs every action, judgment, and interpretation. Without this implicit foundation, practical expertise would collapse under the weight of conscious deliberation, as every pattern, edge, and detail would require explicit thought. True expertise blends cognitive knowledge with embodied practices, resulting in a fluid, intuitive response to complex tasks.

In the high-stakes environment of radiology, where experts analyze complex imaging data in real time, collaborative interpretation of this data becomes an exercise in both collective knowledge and humility. Radiologists and other clinicians, such as surgeons or technicians, gather around the PACS to engage in collaborative decision-making. The team's commitment to shared problem-solving embodies a hermeneutic collaboration that values diverse insights as indispensable assets. In this process, the collective contributions—whether from a radiologist or another member of the clinical team—serve to enrich the team's understanding of the patient's condition. When unexpected issues arise, the team does not merely react; they engage in a deliberate, open-ended dialogue where each voice contributes to a fuller and more nuanced understanding of the situation.

In this interpretive process, "truth" is treated not as a fixed endpoint but as a working hypothesis—one that must be continually refined, challenged, and agreed upon through the team's collaborative inquiry. Through this approach, a unique kind of clinical truth emerges—one that is provisional, contextual, and highly responsive to the specificities of each patient's case. Rather than merely diagnosing or executing a pre-defined plan, the team must constantly adapt, weaving together their diverse perspectives to address the complexities before them. This responsiveness to context and multiplicity requires an openness to differences in interpretation, allowing previously unrecognized biases or assumptions to become known as the team strives for a balanced, comprehensive view.

Importantly, this collaborative process is not aimed at merely reaching consensus; it is *a pursuit of the most accurate understanding possible* in the given moment. This commitment to maintaining an open, interrogative stance enables the team to arrive at solutions that are as nuanced as the clinical challenges they face. By fostering an environment where truth can be questioned and revisited, the team embodies a truly dialogical approach to radiology—one that emphasizes understanding over certainty and embraces diversity of thought.

However, we must recognize that cognitive biases function much like skills—they are ingrained, largely automatic frameworks that shape every aspect of how we think, perceive, and interpret the world around us. Some biases are biological, rooted deeply in human physiology, while others develop through social and cultural influences, much like skills that are part of our tacit knowledge. Ironically, many of these "biases" serve essential purposes. Cognitive shortcuts, such as filling in missing information, blend biological predispositions and cultural learning into almost seamless harmony, enabling us to navigate complex environments with limited information. Yet, these shortcuts do not deliver absolute truths; they generate just enough probable outcomes to allow us to make decisions and take action—even if they are not perfect interpretations of reality.

This suggests that the mind's primary function might not be to create accurate representations but to generate workable interpretations. Our perceptions are not flawless mirrors of the world; they are sometimes interpretative distortions that we accept as "truth" to the extent that they let us function effectively. This process is so ingrained that even if we intellectually recognize these biases as distortions, we still rely on them as practical truths—truths that serve our needs, even if they do not fully reflect the complexity of the world. Thus, cognitive biases are not simply limitations; they are essential tools, finely honed by evolution and culture, which help us navigate an infinitely complex world in ways that are "good enough" to keep us moving forward.

At the heart of interpretation in radiology lies memory—a dynamic, adaptable system that links the pre-conscious and conscious realms. Memory is not a fixed archive of past events but a constantly reconstructing web, shaped by both immediate needs and past experiences. This self-modifying system allows radiologists to connect new experiences to pre-existing frameworks, making sense of images, textures, and spatial cues through implicit patterns rather than explicit reasoning.

The relationship between perception and memory in radiology emerges as a complex, interactive feedback loop. As radiologists engage with images, they do not passively receive the world, but rather recreate it through habitual experience, categorization, and reconnection.

Philosophers like Johnson, Varela, and Merleau-Ponty further enrich this understanding, emphasizing that cognition is not simply located in the mind, but emerges through the embodied, lived experience of interacting with the environment. This perspective finds clear expression in the radiologist's task, where memory and sensory engagement coalesce, allowing for the recognition of patterns without full conscious awareness. Radiologists, like cardiologists and forensic scientists, draw on a "global memory map"—an intricate network of past experience and embodied learning that enables pre-conscious processing. This embodied form of knowing is crucial for making rapid, yet informed decisions based on complex, time-sensitive data.

This notion of memory and cognition as fluid, embodied, and evolving resonates deeply with the neural architecture of the brain. Memory is a vast, multi-layered system capable of adapting and recontextualizing based on new experiences. Edelman and Tononi describe this as a dynamic process that continually recategorizes and reinterprets memories, suggesting that knowledge is not static, but constructive. In radiology, where visual perception plays a critical role, this dynamic restructuring process allows practitioners to adjust their interpretations based on the shifting contexts of each patient's case.

The collaborative, embodied nature of interpretation in radiology reflects a dynamic, evolving process of collective expertise. Memory, perception, and action are not isolated or static, but interwoven in a fluid, ongoing process that shapes how knowledge is constructed in practice. In the collaborative, embodied hermeneutics of radiology, truth is not fixed; it is provisional, contextual, and continually refined through shared inquiry. This approach transforms radiology into an interpretative practice—one in which knowledge is not merely grasped, but actively negotiated, adapted, and reshaped in response to the complexities of each case. The collaborative and embodied nature of radiology practice underscores the profound interdependence of mind, body, and environment in the pursuit of patient care.

In closing, radiological training must evolve beyond the mere acquisition of technical expertise; it should embrace an ongoing interpretative process that fosters deeper engagement with complex cases. While Gadamer highlights the initial moment of understanding, his framework also invites us to consider how interpretation matures through sustained interaction with the subject. In radiology, this means moving beyond static diagnostic criteria and embracing a dynamic learning process informed by continued practice, dialogue, and exposure to diverse cases.

One of the central challenges in radiological training is recognizing that medical imaging does not yield a single, definitive interpretation. Rather, each case presents a range of possibilities that require careful consideration. Trainees must be encouraged to develop an interpretative openness—an ability to reassess initial impressions in light of new information. This plurality of perspectives is not a reflection of subjective disagreement but an acknowledgment of the layered complexity inherent in medical diagnosis.

A useful example of this interpretative process can be seen in the collaborative decision-making among medical professionals. In a case where a senior radiologist suggests a particular diagnosis, their recommendation is not necessarily accepted due to their authority but because it withstands scrutiny within the context of collective expertise. Similar to the way cardiologists engage in dialogue to determine the best course of action for a patient, radiologists must engage in a shared interpretative process that balances clinical knowledge, imaging findings, and patient history.

Gadamer's concept of *Bildung*, or self-formation, is highly relevant in this context. Radiologists do not merely accumulate knowledge; they refine their interpretative sensibilities through experience. This *Bildung* unfolds through the dialectic of experience, wherein repeated exposure to cases fosters a deeper sensitivity to the nuances of radiological interpretation. By consistently engaging with ambiguity, radiologists cultivate an intellectual flexibility that allows them to navigate the inherent uncertainties of medical imaging.

Ultimately, radiological training should emphasize the importance of dialogue, reflective practice, and openness to evolving interpretations. Just as Gadamer posits that reality is inexhaustibly complex, radiologists must accept that definitive understanding is often elusive. Instead of seeking absolute certainty, training should focus on cultivating a readiness to reconsider initial judgments, incorporate emerging insights, and adapt to new medical advancements. Through this interpretative practice, radiologists not only enhance their diagnostic accuracy but also contribute to the collective wisdom of the medical community.

# References

Aalders, M. C., N. L. Adolphi, B. Daly, G. G. Davis, H. H. de Boer, S. J. Decker, J. J. Dempers, J. Ford, C. Y. Gerrard, G. M. Hatch, P. A. M. Hofman, M. Lino, C. Jacobson, W. M. Klein, B. Kubat, P. M. Leth, E. L. Mazuchowski, K. B. Nolte, C. O'Donnell, M. J. Thali, R. R. van Rijn, and K. Wozniak. 2017. Research in Forensic Radiology and Imaging; Identifying the Most Important Issues. *Journal of Forensic Radiology and Imaging* 8:1–8.

Al, E., F. Iliopoulos, N. Forschack, T. Nierhaus, M. Grund, P. Motyka, M. Gaebler, V. V. Nikulin, and A. Villringer. 2020. Heart-Brain Interactions Shape Somatosensory Perception and Evoked Potentials. *PNAS* 117 (19): 10575–10584.

Balding, David. 2015. *Weight-of-Evidence for Forensic DNA Profiles.* Wiley.

Barrett, W. 1956. Zen for the West. In *Selected Writings of D.T. Suzuki—Zen Buddhism*, ed. William Barret. New York: Three Leaves Press, Doubleday.

Barrett, J. F., and N. Keat. 2004. Artifacts in CT Recognition and Avoidance. *Radiographics* 24 (6): 1679–1691.

Barry, A. M. S. 1997. *Visual Intelligence: Perception, Image, and Manipulation in Visual Communication.* Albany: State University of New York Press.

Bassolino, M., A. Serino, S. Ubaldi, and E. Làdavos. 2010. Everyday Use of the Computer Mouse Extends Peripersonal Space Representation. *Neuropsychologia* 48:803–811.

Beaton, M. 2016. Sensorimotor Direct Realism: How We Enact Our World. *Constructivist Foundations* 11 (2): 265–276.

Beck, Jamie J. W. 2011. What is the Future of Imaging in Forensic Practice? *Radiography* 17:212–217.



J. K. B. O. Friis, *Hermeneutics at the Intersection of Medical Technology*, https://doi.org/10.1007/978-3-031-88755-0

Brady, A. P. 2017. Error and Discrepancy in Radiology: Inevitable or Avoidable? *Insights into Imaging* 8 (1): 171–182.

Bruno, M. A., Walker, E. A., Abujudeh, H. H. 2015. Understanding and Confronting Our Mistakes: The Epidemiology of Error in Radiology and Strategies for Error Reduction. *Radiographics* 35 (6): 1668–1676. https://doi.org/10.1148/rg.2015150023. PMID: 26466178.

Buber, M. 1970. *I and Thou*. Translated by Walter Kaufmann. New York: Charles Scribner's Sons.

Bünning, E. 1964. *The Physiological Clock*. Berlin: Springer.

Campbell, J. 1982. *The Grammatical Man*. New York: Simon and Schuster.

Canaris, M. M. 2017. *Francis A. Sullivan, S.J. and Ecclesiological Hermeneutics: An Exercise in Faithful Creativity*. Leiden: Brill NV.

Caputo, J. 1986. *The Mystical Element in Heidegger's Thought*. New York: Fordham University Press.

Casey, E. S. 1984. Habitual Body and Memory in Merleau-Ponty. *Man World* 17:279–297.

Chua, H. F., J. E. Boland, and R. E. Nisbett. 2005. Cultural Variation in Eye Movements during Scene Perception. *Proceedings of the National Academy of Science* 102 (35): 12629–12633.

Clark, A. 2015. Chapter 1: Prediction Machines. In *Surfing Uncertainty: Prediction, Action, and the Embodied Mind*, 12–15. https://academic.oup.com.

Clark, A. 2016. *Surfing Uncertainty: Prediction, Action, and the Embodied Mind*. Oxford University Press.

Clark, A., and D. Chalmers. 1998. The Extended Mind. *Analysis* 58 (1): 7–19.

Conway, F., and J. Siegelman. 2005. *Dark Hero of The Information Age—I Search of Norbert Wiener the Father of Cybernetics*. Basic Books.

Craighero, L. 2014. The Role of the Motor System in Cognitive Functions. In *The Routledge Handbook of Embodied Cognition*, ed. L. A. Shapiro. Routledge.

Damasio, A. 1994. *Descartes Error: Emotion, Reason, and the Human Brain*. New York: G.P. Putnam.

Davey, N. 1999. The Hermeneutic of Seeing. In *Interpreting Visual Culture. Explorations in the Hermeneutics of the Visual*, ed. I. Heywood and B. Sandywell. Routledge.

de Boer, B., and J. Zwier, eds. 2024. *Phenomenology and the Philosophy of Technology*. Open Book Publishers.

De Yoe, E., and D. C. Van-Essen. 1988. Concurrent Processing Streams in Monkey Visual Cortex. *Trends in Neurosciences* 11 (5): 219–226.

Degenaar, J., and J. K. O'Regan. 2015. Sensorimotor Theory of Consciousness. *Scholarpedia* 10 (5): 4952.

Dennett, D., and M. Kinsbourne. 1992. Time and the Observer. *Behavioral and Brain Sciences* 15:183–247.

Di Paolo, E. A., and E. Thompson. 2014. The Enactive Approach. In *The Routledge Handbook of Embodied Cognition*, ed. L. A. Shapiro. Routledge.

Di Paolo, E. A., M. Rhohde, and H. De Jaegher. 2014. Horizons for the Enactive Mind: Values, Social Interaction, and Play. In *Enaction: Toward a New Paradigm for Cognitive Science*, ed. John Stewart, Oliver Gapenne, and Ezequiel A. Di Paolo. MIT Press.

Dockrill, P. 2020. Every Single Heartbeat Affects Our Perception of The World, Researchers Find. Sciencealert.com (Edition May 1, 2020).

Donald, J. J., and S. A. Barnard. 2012. Common Patterns in 558 Diagnostic Radiology Errors. *Journal of Medical Imaging and Radiation Oncology* 56 (2): 173–178.

Doxiadis, A., and C. H. Papadimitriou. 2009. *Logicomix. An Epic Search for Truth.* Bloomsbury.

Dreyfus, H. L., and S. E. Dreyfus. 1986. *Mind over Machine. The Power of Human Intuition and Expertise in the Era of Computer.* With Tom Athanasiou. Basil Blackwell.

Du, M. 2017. Analysis of Errors in Forensic Science. *Journal of Forensic Science and Medicine* 3 (3): 139–143. Wolters Kluwer—Medknow.

Eddington, A. 1946. *The Nature of the Physical World.* Cambridge: Cambridge University Press.

Eddington, A. 1976. The Arrow of Time, Entropy and The Expansion of the Universe. In *The Concepts of Space and Time*, BSPS, ed. M. Capek, vol. XXII. D: Reidel Publishing Company.

Edelman, G., and G. Tononi. 2000. *A Universe of Consciousness: How Matter Becomes Imagination.* Basic Books.

Edmond, G., A. Towler, B. Growns, G. Ribeiro, B. Found, D. White, K. Ballantyne, R. A. Searston, M. B. Thompson, J. M. Tangen, R. I. Kemp, and K. Martire. 2017. Thinking Forensics: Cognitive Science for Forensic Practitioners. *Science and Justice* 57:144–154.

Enns, J. T. 2004. *The Thinking Eye, the Seeing Brain: Explorations in Visual Cognition.* New York: Norton.

Érdi, P. 1983. Hierarchical Thermodynamic Approach to the Brain. *The International Journal of Neuroscience* 20:193–216.

Érdi, P. 1996. The Brain as a Hermeneutic Device. *Biosystems* 38:179–189.

Érdi, P. 1999. Neural and Mental Development: Selectionism, Constructivism, Hermeneutics. In *Neuronal Bases and Psychological Aspects of Consciousness*, ed. C. Taddei-Ferrati, 507–518. World Scientific.

Érdi, P., and Tsuda, I. 2002. Hermeneutic Approach to the Brain: Process versus Device? *Theoria et Historia Scientiarum* VI (2).

Eriksson, Anders, Torfinn Gustafsson, Malin Höistad, Monica Hultcrantz, Stella Jacobson, Ingegerd Mejare, and Anders Persson. 2017. Diagnostic Accuracy of

Postmortem Imaging vs Autopsy—A Systematic Review. *European Journal of Radiology* 89:249.

Figal, G. 2010. *Objectivity—The Hermeneutical and Philosophy*. Translated by Theodore D. George. SUNY Press.

Fishbein, A. B., K. L. Knutson, and P. C. Zee. 2021. Circadian Disruption and Human Health. *Journal of Clinical Investigation* 131 (19): e148286.

Fløistad, G. 1987. Selvransakelse. In *Tanke og Mistanke*, ed. Asbjørn Aarnes and Helge Salemonsen. Oslo: Aventura Forlag.

Føllesdal, D. 1979. Hermeneutics and the Hypothetico-Deductive Method. *Dialectica* 3 (3–4): 319–336.

Forrest, P. 2010. The Identity of Indiscernibles. In *Stanford Encyclopedia of Philosophy*. https://plato.stanford.edu/entries/identity-indiscernible/.

Frank, J. M., R. C. O'Reilly, and T. Curran. 2006. When Memory Fails. Intuition Reigns. Midazolam Enhances Implicit Inference in Humans. *Psychological Science* 17 (8): 700–707.

Fraser, J. T. 1990. *Of Time, Passion, and Knowledge*. Princeton University Press.

Freeman, W. 1991. The Physiology of Perception. *Scientific American* 264:78–85.

Friis, J. K. B. O. 2011a. Interpreting the Visual. *Philosophy and Technology* 25 (2): 249–270.

Friis, J. K. B. O. 2011b. Perception: Embodiment and Beyond. In *Opening Up the In-between: Interdisciplinary Reflections on Science, Technology and Social Change*. Foundations of Science, Springer.

Friis, J. K. B. O. 2015a. Towards a Hermeneutics of Unveiling. In *Postphenomenology*, ed. Robert Rosenberger, Peter-Paul Verbeek, and Don Ihde. Lexington Books.

Friis, J. K. B. O. 2015b. Gestalt Descriptions Embodiments and Medical Image Interpretation. *AI & Society* 32:209–218.

Friis, J. K. B. O. 2016a. The Interpretative Value of Information. In *The Routledge Handbook of Philosophy of Information*, ed. L. Floridi. Routledge.

Friis, J. K. B. O. 2016b. Tacitly Thinking. *Quadranti – Rivista Internazionale di Filosofia Contemporana*. ISSN 2282–4219.

Friis, J. K. B. O. 2019. Philosophy of Technology. In *The Cambridge History of Philosophy, 1945–2015*, ed. K. Becker and I. Thomson, 497–512. Cambridge: Cambridge University Press.

Friis, J. K. B. O. 2020a. Enactive Hermeneutics and Smart Medical Technologies. *AI & Society* 38:2141–2149. Springer.

Friis, J. K. B. O. 2020b. Review of Ian Heywood and Barry Sandywell (eds.). Interpreting Visual Culture. Explorations in the Hermeneutics of the Visual. *AI & Society* 38:2369–2373.

Friis, J. K. B. O. 2021. Radiology as Skillful Coping and Enactive Hermeneutics: A Critique of Representations and Corresponding Truth. In *Postphenomenology and Imaging: How to Read Technology*, ed. Samantha Jo Fried and Robert Rosenberger. Lexington Books. Forthcoming.

Friston, K. J., and C. D. Frith. 2015. Active Inference, Communication and Hermeneutics. *Cortex: A Journal Devoted to the Study of the Nervous System and Behavior* 68:129–143. https://doi.org/10.1016/j.cortex.2015.03.025.

Gabrielson, Ryan. 2019. The FBI Says Its Photo Analysis Is Scientific Evidence. Scientists Disagree. *ProPublica*.

Gadamer, H.-G. 1976. *Philosophical Hermeneutics*. Translated by D. E. Linge. Berkeley and Los Angeles, CA: University of California Press.

Gadamer, H.-G. 2007. *Sandhed og Metode*. Aarhus: Academica.

Gadamer, H.-G., ed. 2013. *Truth and Method*. Bloomsbury.

Gallagher, S. 2004. Hermeneutics and the Cognitive Sciences. *Journal of Consciousness Studies* 11:10–11.

Gallagher, S. 2005. *How the Body Shapes the Mind*. Oxford: Oxford University Press.

Gallagher, S. 2008. Intersubjectivity in Perception. *Continental Philosophy Review* 41 (2): 163–178.

Gallagher, S. 2011a. Interpretations of Embodied Cognition. In *The Implications of Embodiment: Cognition and Communication*, ed. W. Tschacher and C. Bergomi. Exeter: Imprint.

Gallagher, S. 2011b. *The Overextended Mind*. Versus: Quaderni di studi semiotici.

Gallagher, S., and M. Allen. 2018. Active Inference, Enactivism and the Hermeneutics of Social Cognition. *Synthese* 195:2627–2648.

George, T. D. 2009. Günter Figal's Hermeneutics. *Philosophy Compass* 4 (6): 904–912.

George, T. D. 2020. Hermeneutics. In *Stanford Encyclopedia of Philosophy*.

George, T. D. 2022. Is Hermeneutics Realistic? On the Normative Orientation toward Plurality. *Critical Hermeneutics* 6 (2): 191–210. Biannual International Journal of Philosophy.

Gibson, J. J. 1979. *The Ecological Approach to Visual Perception*. Houghton Mifflin.

Giger, M. L. 2018. Machine Learning in Medical Imaging. *Journal of the American College of Radiology* 15 (3): 512–520.

Gill, K. S. 2018. Making Sense: Cognition, Art, and Embodiment. *AI & Society*. Springer.

Glück, L. 1968. *Firstborn*. Ecco Press.

Gödel, K. 1931. On Formally Undecidable Propositions of Principia Mathematica and Related Systems I. In *2000. Engines of Logic: Mathematics and the Origin of the Computer*, ed. Martin Davis. New York: W. W. Norton & Company.

Goyal, B., A. Dogra, S. Agrawal, and B. S. Sohi. 2018. Noise Issues Prevailing in Various Types of Medical Images. *Biomedical and Pharmacology Journal* 11 (3): 1227.

Graber, M. L., N. Franklin, and R. Gordon. 2014. Diagnostic Error in Internal Medicine. *Archives of Internal Medicine* 165 (13): 1493–1499.

Green, M. 2018. Errors in Eyewitness Identification Procedures. https://www.visualexpert.com/Resources/mistakenid.html. Last visited 20 May 2023.

Gregory, R. L. 1994. Putting Illusions in Their Place. In *Even Odder Perceptions.* London: Routledge and Kegan Paul.

Grondin, J. 1993. *L'universalité de l'herméneutique.* Paris: PUF.

Grondin, J. 2005. Hermeneutics. In *New Dictionary of the History of Ideas*, ed. M. C. Horowitz, 982–987.

Grondin, J. 2022. Foreword to the Arabic translation of *Truth and Method.* In *Wahrheit und Methode. Grundzüge einer philosophischen Hermeneutik* (in Arabic), ed. Hans-Georg Gadamer. Oeabooks.

Grush, R. 2006. How to, and How *Not* to, Bridge Computational Cognitive Neuroscience and Husserlian Phenomenology of Time Consciousness. *Synthese* 153:417–450.

Gudbjartsson, H., and S. Patz. 1995. The Rician Distribution of Noisy MRI Data. *Magnetic Resonance in Medicine* 34 (6): 910–914.

Guitchounts, G. 2020. An Existential Crisis in Neuroscience: We're mapping the brain in amazing detail–but our brain can't understand the picture. *Nautilus.* http://nautil.us/issue/81/maps/an-existential-crisis-in-neuroscience. Accessed 23 January 2020.

Hald, C. 2014. Philosophical Basis of the Forensic Process. In *Encyclopedia of Criminology and Criminal Justice*, ed. G. Bruinsma and D. Weisburd. New York, NY: Springer. https://doi.org/10.1007/978-1-4614-5690-2_380.

Haldar, P. 2013. Forensic Representations of Identity: The *Imago*, the X-Ray and the Evidential Image. *Law and Humanities* 7 (2): 129.

Hanson, N. R. 1958. *Patterns of Discovery: An Inquiry into the Conceptual Foundations of Science.* Cambridge: Cambridge University Press.

Harris, E. E. 1988. *The Reality of Time.* SUNY.

Harris, E. E. 1993. *The Foundations of Metaphysics in Science.* New Jersey and London: Humanities Press.

Hasse, C. 2020. *Posthumanist Learning. What Robots and Cyborgs Teach us About Being Ultra-social.* Routledge.

Heckers, S., M. Zalesak, A. P. Weiss, and T. Ditman. 2004. Hippocampal Activation During Transitive Inference in Humans. *Hippocampus* 14:153–162.

Heidegger, M. 1966. *Discourse on Thinking: A Translation of Gelassenheit.* Translated by J. Anderson and E. Freund. Harper & Row.

Heidegger, M. 1992a. *Being and Time.* Translated by John Macquarrie and Edward Robinson. Oxford: Blackwell.

Heidegger, M. 1992b. *History of the Concept of Time: Prolegomena.* Bloomington, IN: Indiana University Press.

Helmholtz, H. 1867. *Treatise on Physiological Optics.* Dover.

Heywood, I. 1999. "Ever More Specific"—Practices and Perceptions in Art and Ethics. In *Interpreting Visual Culture. Explorations in the Hermeneutics of the Visual*, ed. I. Heywood and B. Sandywell. Routledge.

Heywood, I., and B. Sandywell, eds. 1999. *Interpreting Visual Culture. Explorations in the Hermeneutics of the Visual.* Routledge.

Hodge, S. 2018. Current Controversies in the Use of DNA in Forensic Investigations. *University of Baltimore Law Review* 48 (1): 3.

Hohwy, J. 2016. The Self-Evidencing Brain. *Noûs* 50 (2): 259–285.

Hume, D. 1990. *Enquiries Concerning Human Understanding and Concerning the Principles if Morals.* Reprinted from the 1777 edition. Revised by P. H. Nidditch. Oxford University Press.

Husbands, P., T. Smith, N. Jakobi, and M. O'Shea. 1998. Better Living through Chemistry: Evolving GasNets for Robot Control. *Connection Science* 10:185–210.

Hutto, D. D. 2012. Truly Enactive Emotion. *Emotion Review* 4 (2): 1–6.

Hutto, D. D. 2013. Enactivism, From A Wittgensteinian Point of View. *American Philosophical Quarterly* 50 (3): 281–302.

Ihde, D. 1998. *Expanding Hermeneutics. Visualism in Science.* Evanston, IL: Northwestern University Press.

Ihde, D. 2009. *Postphenomenology and Technoscience—The Peking University Lectures.* New York: SUNY Press.

Ihde, D. 2011. Stretching the In-between: Embodiment and Beyond. *Foundations of Science* 16 (2–3): 109–118.

Ihde, D. 2015. Preface: Positioning Postphenomenology. In *Postphenomenological Investigations: Essays on Human-Technology Relations*, ed. Robert Rosenberger and Peter-Paul Verbeek. Lexington Books.

Ingarden, R. 1970. *Innføring i Edmund Husserls fenomenologi: 10 Oslo forelesninger.* Oslo: J.G. Tanum.

Itri, J. N., and S. H. Patel. 2018. Heuristics and Cognitive Error in Medical Imaging. *American Journal of Roentgenology* 210 (5): 1095–1105.

Jagannath, A., L. Taylor, Z. Wakaf, S. R. Vasudevan, and R. G. Foster. 2017. The Genetics of Circadian Rhythms, Sleep and Health. *Human Molecular Genetics* 26 (R2): R128–R138.

Jamieson, A. 2008. The Philosophy of Forensic Scientific Identification. *Hastings Law Journal* 59:1031.

Jerison, H. J. 1973. *Evolution of the Brain and Intelligence.* Academic Press.

Jerison, H. J. 1976. Paleoneurology and the Evolution of Mind. *Scientific American* 234 (1): 90–101.

Johnson, M. 2017. *Embodied Mind, Meaning, and Reason—How Our Bodies Give Rise To Understanding.* The University of Chicago Press.

Kavitha, B., A. Einstein, B. Sivapathasundharam, and T. R. Saraswathi 2009. Limitations in Forensic Odontology. *Journal of Forensic Dental Sciences* 1 (1).

Koontz, B. H., and R. B. Gunderman. 2008. Gestalt Theory: Implications for Radiology Education. *AJR American Journal of Roentgenology* 190 (5): 1156–1160.

Krupinski, E. A. 2011. The Role of Perception in Imaging: Past and Future. *Seminars in Nuclear Medicine* 41 (6): 392–400.

Krupinski, E. A., H. L. Kundel, P. F. Judy, and C. F. Nodine. 1998. Key Issues for Image Perception Research. *Radiology* 209:611–612.

Kundel, H. L. 1990. Visual Cues in the Interpretation of Medical Images. *Journal of Clinical Neurophysiology* 7 (4): 472–483.

Kundel, H. L. 2006. History of Research in Medical Image Perception. *American College of Radiology* 3:402–408.

Kundel, H. L., et al. 2007. Cognitive Errors in Radiology. *Academic Radiology* 14 (7): 782–786.

Lahiri, A. 2020. *Inference Belief and Interpretation in Science.* Website: physicsandmore.net // physics-plus-more.blogspot.com.

Laney, C., and E. F. Loftus. 2008. Emotional Content of True and False Memories. *Memory* 16 (5): 500–516.

Leone, M. 2021. From Fingers to Faces: Visual Semiotics and Digital Forensics. *International Journal for the Semiotics of Law* 34:579–599.

Lestienne, R. 1995. *The Children of Time.* University of Illinois Press.

Lestienne, R. 1998. Time, Chaos, and Memory: From Physical Objects to Living Beings. In *Time, Order, Chaos, The Study of Time IX*, ed. J. T. Fraser, M. P. Soulsby, and A. Argyros. International Universities Press, Inc.

Lev, M., J. T. Rhea, and R. T. Bramson. 1999. Avoidance of Variability and Error in Radiology. *The Lancet* 354 (9175): P272.

Levin, D. M. 1999. My Philosophical Project and the Empty Jug. In *Interpreting Visual Culture. Explorations in the Hermeneutics of the Visual*, ed. I. Heywood and B. Sandywell. Routledge.

Livingston, P. 2005. Functionalism and Logical Analysis. In *Phenomenology and Philosophy of Mind*, ed. D. Woodruff Smith and A. L. Thomasson. Oxford University Press.

Loftus, E. F. 1992. When a Lie Becomes Memory's Truth: Memory Distortion After Exposure to Misinformation. *Current Directions in Psychological Science* 1 (4): 121–123.

Magnusson, T. 2009. Of Epistemic Tools: Musical Instruments as Cognitive Extensions. *Organised Sound* 14 (2): 168–176.

Manning, D. J., A. Gale, and E. A. Krupinski. 2005. Perception Research in Medical Imaging. *British Journal of Radiology* 78:683–685.

Mantzavinos, C. 2005. *Naturalistic Hermeneutics.* Cambridge: Cambridge University Press.

Marr, D. 2001. The Philosophy and the Approach. In *Key Readings in Cognition: Visual Perception*, ed. S. Yantis. New York: Psychology Press.

Maturana, H., and F. Varela. 1980. *Autopoiesis and Cognition. The Realization of the Livings*. D. Reidel Publishing Company.

Maturana, H., and F. Varela. 1987. *The Tree of Knowledge. The Biological Roots of Human Understanding*. Shambhala Publications.

Mejer, M. 2022. *Updating the Interpretative Turn: New Arguments in Hermeneutics*. Routledge.

Mercier, P. 2007. *Night Train to Lisbon*. Grove Press.

Merleau-Ponty, M. 1962. *Phenomenology of Perception*. Translated by Colin Smith. Routledge Classics. Reprinted version 2003.

Merleau-Ponty, M. 1983. *The Structure of Behavior*. Translated by Alden L. Fisher. Duquesne University Press.

Merton, T. 1968. *Zen and the Birds of Appetite*. New Directions.

Monk, R. 2000. *Bertrand Russell—The Ghost of Madness 1921–1970*. Vintage.

Montaruli, A., L. Castelli, A. Mulè, R. Scurati, F. Esposito, L. Galasso, and E. Roveda. 2021. Biological Rhythm and Chronotype: New Perspectives in Health. *Biomolecules* 11 (4): 487.

Morrison, G. S. 2022. Advancing a Paradigm Shift in Evaluation of Forensic Evidence: The Rise of Forensic Data Science. *Forensic Science International: Synergy* 5:100270.

Musil, R. 1995. *The Man Without Qualities*. Picador.

Nietzsche, F. 2011. *Thus Spoke Zarathustra*. Simon & Brown.

Noë, A. 2004. *Action in Perception*. Cambridge: The MIT Press.

Nonaka, I., and I. Yamaguchi. 2022. *A Dynamic Management Theory Predicated on the "Philosophy of Empathy"*. Palgrave Macmillan.

Nowotny, H. 1994. *Time—The Modern and Postmodern Experience*. Polity Press.

Ovens, M. 2013. The Use of Hermeneutics in the Criminological Interpretation of Forensic Documents: A Case Study. *Journal of Psychology in Africa* 23 (4): 661.

Patke, A., M. W. Young, and S. Axelrod. 2020. Molecular Mechanisms and Physiological Importance of Circadian Rhythms. *Nature Reviews Molecular Cell Biology* 21 (2): 67–84.

Pickering, J. 2016. Who is "We"? Some Observations on Sensorimotor Direct Realism. *Constructivist Foundations* 11 (2): 279–280. Commentary on Michael Beaton's Sensorimotor Direct Realism: How We Enact Our World.

Pittendrigh, C. S. 1981. Circadian Organization and the Photoperiodic Phenomena. In *Biological Clocks in Seasonal Reproductive Cycles*, ed. B. K. Follett and D. E. Follett, 1–35. Bristol: Wright.

Rasmussen, J., and A. Jensen. 1974. Mental Procedures in Real-Life Tasks: A Case Study of Electronic Trouble Shooting. *Ergometrics* 17 (3): 293–307.

Ricoeur, P. 1974. *The Conflict of Interpretations*. Chicago, IL: Northwestern University Press.

Rosch, E., C. B. Mervis, W. D. Gray, M. Johnson, and P. Boyes-Bream. 1976. Basic Objects in Natural Categories. *Cognitive Psychology* 8:382–439.

Rosenberger, R., and P. P. Verbeek. 2015. A Field Guide to Postphenomenology. In *Postphenomenological Investigations: Essays on Human-Technology Relations*, ed. R. Rosenberger and P. P. Verbeek. Lexington Books.

Ruin, H. 2019a. Looking into the Open—Nietzsche and the New Quest for the Real. In *Die Gegenständlichkeit der Welt. Festschrift für Günter Figal zum 70. Geburtstag*, ed. A. Egel, D. Espinet, T. Keiling, and B. Zimmermann. Mohr Siebeck Tübingen.

Ruin, H. 2019b. The Inversion of Mysticism—*Gelassenheit* and the Secret of the Open in Heidegger. *Religions* 10:15.

Sabih, D., et al. 2010. Image Perception and Interpretation of Abnormalities: Can We Believe Our Eyes? Can We Do Something About It? *European Society of Radiology* 1–2.

Sabih, D., A. Sabih, Q. Sabih, and A. N. Khan. 2011. Image Perception and Interpretation of Abnormalities; Can We Believe Our Eyes? Can We Do Something About It? *Insights into Imaging* 2 (1): 47–55.

Saks, M. J., and J. J. Koehler. 2005. The Coming Paradigm Shift in Forensic Identification Science. *Science* 309:892–895.

Saks, M. J., and J. J. Koehler. 2019. The Individualization Fallacy in Forensic Science Evidence. *Vanderbilt Law Review* 61 (1): 199–219.

Sandywell, B. 1999. Specular Grammar: The Visual Rhetoric of Modernity. In *Interpreting Visual Culture. Explorations in the Hermeneutics of the Visual*, ed. I. Heywood and B. Sandywell. Routledge.

Sandywell, B. 2011. *Dictionary of Visual Discourse. A Dialectical Lexicon of Terms.* Routledge.

Saunders, D. S. 1968. Photoperiodism and Time Measurement in the Parasitic Wasp, Nasonia vitripennis. *Journal of Insect Physiology* 1968 (14): 433–450.

Saunders, D. S. 1974. Evidence for 'Dawn' and 'Dusk' Oscillators in the Nasonia Photoperiodic Clock. *Journal of Insect Physiology* 20:77–88.

Saunders, D. S. 1976. The Biological Clocks of Insects. *Scientific American* 234 (2): 114–121.

Saunders, D. S. 2002. *Insect Clocks.* Amsterdam: Elsevier Science.

Schneider, P. M., B. Prainsack, and M. Kayser. 2019. The Use of Forensic DNA Phenotyping in Predicting Appearance and Biogeographic Ancestry. *Deutsches Arzteblatt International* 116:873–880.

Shapiro, L. A. 2004. *The Mind Incarnate.* Cambridge, MA: MIT Press.

Shapiro, L. A., ed. 2014. *The Routledge Handbook of Embodied Cognition.* Routledge.

Sklar, L. 1995. Time in Experience and in Theoretical Description of the World. In *Time's Arrows Today*, ed. S. F. Savitt. Cambridge: Cambridge University Press.

Snow, C. P. 1963/1993. *The Two Cultures*. Cambridge University Press.
Snow, J. C., Johnson, E. L., & Culham, J. C. 2011. Repetition effects in object-selective cortex are modulated by stimulus orientation. *Journal of Cognitive Neuroscience* 23 (7): 1685–1698. https://doi.org/10.1162/jocn.2010.21527
Sonsino, S. 2005. Towards a Hermeneutics of Narrative Identity: A Ricoeurian Framework for Exploring Narratives (and Narrators) of Strategy. *Organization Management Journal* 2 (3): 166–182.
Stoney, D. A. 1991. What Made Us Ever Think We Could Individualize Using Statistics? *Journal of the Forensic Science Society* 31 (2): 197–199.
Szondi, P. 1995. *Introduction to Literary Hermeneutics*. Cambridge University Press.
Tarkovsky, A. 1989. *Sculpting in Time: Reflections on the Cinema*. University of Texas Press.
Thompson, E. 2007. *Mind in Life: Biology, Phenomenology, and the Sciences of Mind*. Cambridge, MA: Harvard University Press.
Tsuda, I. 1984. A Hermeneutic Process of the Brain. *Progress of Theoretical Physics Supplement* 79.
Tsuda, I. 1991. Chaotic Itineracy as a Dynamical Basis of Hermeneutic Sin Brain and Mind. *World Futures* 32:167–184.
Tsuda, I. 1993. Chaotic Hermeneutics for Understanding the Brain. In *Endophysics: The World from Within. A New Approach to the Observer-Problem with Applications in Physics, Biology and Mathematics*, ed. George Kampis, 219–234. Santa Cruz: Published by Aerial. ISBN 978-0685704943.
Turing, A. M. 1950 (1990). Computing Machinery and Intelligence. *Mind, 59*. In *The Philosophy of Artificial Intelligence*, ed. M. A. Boden, 40–66. Oxford: Oxford University Press.
Van Fraassen, B. C. 2002. *The Empirical Stance*. Yale University Press.
Varela, F. 1999. *Ethical Know-How: Action, Wisdom, and Cognition*. Stanford University Press.
Varela, F. J., E. Thompson, and E. Rosch. 1993. *The Embodied Mind: Cognitive Science and Human Experience*. MIT Press.
Verbeek, P.-P. 2003. Material Hermeneutics. *Techné: Research in Philosophy and Technology* 6 (3): 181–184.
Verbeek, P.-P. 2008. Cyborg Intentionality: Rethinking the Phenomenology of Human-Technology Relations. *Phenomenology and the Cognitive Sciences* 7:387–395.
Virtual Labs. 2023. Identification of Finger Prints Patterns. *Vlabs.iitb.ac.in*. https://da-iitg.vlabs.ac.in/exp/fingerprint-patterns/theory.html. Accessed 30 June 2023.

Vogeley, K., and C. Kupke. 2007. Disturbances of Time Consciousness From a Phenomenological and a Neuroscientific Perspective. *Schizophrenia Bulletin* 33 (1): 157–165.

Wheeler, M. 2014. Revolution, Reform, or Business as Usual? The Future Prospects for Embodied Cognition. In *The Routledge Handbook of Embodied Cognition*, ed. L. Shapiro. Routledge.

Wittgenstein, L. 1989. *Tractatus Logico-Philosophicus*. Routledge.

Yantis, S., ed. 2001. *Key Readings in Cognition: Visual Perception*. New York: Psychology Press. From Yantis' Introduction to Helmholtz' "Concerning the Perception in General".

Zhenhua, Yu. 2003. Tacit Knowledge/Knowing and the Problem of Articulation. *Tradition and Discovery* 30 (2): 11–23.

# Index




J. Friis, *Hermeneutics at the Intersection of Medical Technology*,
https://doi.org/10.1007/978-3-031-88755-0

The manufacturer's authorised representative in the EU is Springer Nature Customer Service Centre GmbH, Europaplatz 3, 69115 Heidelberg, Germany. If you have any concerns regarding our products, please contact ProductSafety@springernature.com

Printed and bound by CPI Group (UK) Ltd, Croydon, CR0 4YY
30/09/2026
02225583-0018